The Chemistry
Perfumes, Pigmen

Margareta Séquin
*Department of Chemistry and Biochemistry,
San Francisco State University, San Francisco, USA
E-mail: msequin@sfsu.edu*

RSCPublishing

ISBN: 978-1-84973-334-2

A catalogue record for this book is available from the British Library

© Margareta Séquin 2012

All rights reserved

Apart from fair dealing for the purposes of research for non-commercial purposes or for private study, criticism or review, as permitted under the Copyright, Designs and Patents Act 1988 and the Copyright and Related Rights Regulations 2003, this publication may not be reproduced, stored or transmitted, in any form or by any means, without the prior permission in writing of The Royal Society of Chemistry or the copyright owner, or in the case of reproduction in accordance with the terms of licences issued by the Copyright Licensing Agency in the UK, or in accordance with the terms of the licences issued by the appropriate Reproduction Rights Organization outside the UK. Enquiries concerning reproduction outside the terms stated here should be sent to The Royal Society of Chemistry at the address printed on this page.

The RSC is not responsible for individual opinions expressed in this work.

Published by The Royal Society of Chemistry,
Thomas Graham House, Science Park, Milton Road,
Cambridge CB4 0WF, UK

Registered Charity Number 207890

For further information see our web site at www.rsc.org

The Chemistry of Plants
Perfumes, Pigments, and Poisons

Preface

Living plants capture our attention with their bright colors, with the fragrances of their flowers, and with their uses as food plants or as medicinal herbs. The subject of chemistry, on the other hand, tends to invoke thoughts of abstract, difficult to understand concepts and is often seen as disconnected from the living world. This book intends to ease access to chemistry, particularly organic chemistry, by combining it with examples from the world of plants.

Everyday observations of the natural world may bring on questions such as "Why do leaves change color in the fall?", "What makes roses smell sweet?", "What makes pine cones sticky?" and "What adds the sharp taste to mustard seeds?" These questions—and many more—are answered in this book by providing an understandable introduction to organic chemistry. Many texts on plant chemistry provide in-depth information that requires considerable knowledge of chemistry. Others provide information in a narrative style, with few chemical structures. This book takes an in-between approach. After an introduction to chemistry concepts as they relate to plants, the book progresses to the fascinating organic structures of plant scents, pigments, and toxins and then concludes with human uses of plant substances.

The experiences from teaching a popular undergraduate college level course to non-chemistry majors have provided the materials for this book. The author has also taught numerous field seminars

---

The Chemistry of Plants: Perfumes, Pigments, and Poisons
Margareta Séquin
© Margareta Séquin 2012
Published by the Royal Society of Chemistry, www.rsc.org

in plant chemistry, at botanical gardens and at national parks, addressing general audiences who were interested in plants. The questions and comments from students, and topics that people were particularly interested in, have greatly contributed to the content of this book.

The text is designed to be useful both to readers who proceed chapter by chapter, and to those who decide to select chapters of special interest. Formulas and chemical structures may look daunting at first. To make the chemistry more accessible, numerous photos of characteristic plants accompany the chemical structures.

Many people have supported me in the preparation of this book, and their help is acknowledged with deep gratitude. Joan Hamilton provided encouraging and most knowledgeable assistance to get the book project started. Prof. Marc Anderson generously provided instructions on the drawing of figures. I am most grateful to Prof. Urs Séquin, Dr. Eileen Nottoli, and Verena Rau who critically read and re-read the entire book, provided advice on content, commented on clearness, and scanned for errors. Stimulating discussions with faculty members of the Chemistry and Biochemistry Department at San Francisco State University about the book's topics provided ideas and helpful insights. Plant materials for photography, with expert plant advice and checking of plant names, were generously provided by Martin Grantham, Greenhouse Curator at San Francisco State University. The contributions of plants and helpful recommendations by staff at the University of California Botanical Garden in Berkeley, and the Regional Parks Botanic Garden in Berkeley, are gratefully acknowledged. Many thanks also go to my friends with great gardens who generously offered the bounty of their plant crops for taking photographs. I am very grateful for the interest and the enthusiasm of students and field-trip audiences that encouraged the writing of this book. My special thanks go to Janet Freshwater, Alice Toby-Brant, and the staff at the Royal Society of Chemistry whose encouragement, support, and expert editorial assistance made this book possible.

Last but not least, I am grateful to my husband Carlo for his patience and support, and for providing cheerful company on photo expeditions.

Contents

Chapter 1
Basic Plant Chemistry Concepts   **1**

1.1 Introduction   1
1.2 Plants and their Elements   3
     1.2.1 Elements and Atoms   9
     1.2.2 Chemical Bonding to Form Molecules   13
1.3 Life-Giving Water   15
1.4 Nutrients from the Soil   23
1.5 Plants and Chemical Reactions   27
1.6 Photosynthesis, Key to Life   28
     1.6.1 Light Reactions and Calvin Cycle   30
     1.6.2 Special Mechanisms in Plants from Hot Climates   31
     1.6.3 Chlorophyll *a*, Main Photosynthetic Pigment   33
1.7 Respiration to Provide Energy   36
1.8 Organic Molecules, an Introduction   39
1.9 Conclusion   45
Bibliography and Further Reading   45
References   46

Chapter 2
The Molecular Building Blocks   **48**

2.1 Introduction   48
2.2 Carbohydrates for Energy and Structure   49
2.3 Structures that Keep Plants Upright   56

The Chemistry of Plants: Perfumes, Pigments, and Poisons
Margareta Séquin
© Margareta Séquin 2012
Published by the Royal Society of Chemistry, www.rsc.org

2.4	Fats and Oils for Energy and Protection	59
2.5	Proteins with Many Functions	65
	2.5.1 Amino Acids, the Building Blocks of Proteins	65
	2.5.2 Peptides and Proteins	68
2.6	Nucleic Acids and Genetic Information	70
2.7	Reaction Pathways that Link Plant Molecules	74
2.8	Conclusion	76
References		77

Chapter 3
Perfumes, Fragrant or Foul 79

3.1	Introduction	79
3.2	Fragrant Terpenes	81
3.3	Sweet Vanilla and Other Aromatics	84
3.4	Pleasant-Smelling Esters	85
3.5	Malodorous Amines and Sulfurous Compounds	86
3.6	Analyzing Plant Odors	88
3.7	Conclusion	90
References		91

Chapter 4
Colorful Plant Pigments 93

4.1	Introduction	93
4.2	The Chlorophylls	95
4.3	Yellow, Orange, and Red Carotenoids	97
4.4	White and Pale-Yellow Flavones	100
4.5	Purple, Pink, and Blue Anthocyanins	102
4.6	Purple Betalains	105
4.7	Brown Tannins	108
4.8	Conclusion	112
References		112

Chapter 5
Poisons and Other Defenses 114

5.1	Introduction	114
5.2	Strong Scents in Leaves	116
5.3	Sticky Resins	119

5.4	Defensive Sulfurous Smells and Tastes	121
5.5	Sour Acids	123
5.6	Bitter Tastes	126
5.7	Milky Saps	128
5.8	Irritants	130
5.9	Growth-Repressing Allelopaths	132
5.10	Harmful Cyanides	133
5.11	Soapy Saponins	135
5.12	Defensive Cardiac Glycosides	137
5.13	Potent Alkaloids	140
5.14	Conclusion	142
References		143

Chapter 6
Plants and People — 145

6.1	Introduction	145
6.2	Foods from Plants	146
	6.2.1 Essential Primary Metabolites	146
	6.2.2 Vitamins	148
	6.2.3 Flavors, Herbs, and Spices	152
6.3	Plant Medicines	155
	6.3.1 History and Introduction	155
	6.3.2 Classic Medicines from Plants	157
	6.3.3 Newer Discoveries	161
6.4	Psychoactive Plants	163
6.5	Fibers and Dyes from Plants	168
	6.5.1 Plant Fibers	168
	6.5.2 Plant Dyes, an Introduction	171
	6.5.3 Classic Plant Dyes	172
	6.5.4 Plant Dyes and Contemporary Colorants	175
6.6	Perfumes for People	176
	6.6.1 Introduction and History	176
	6.6.2 Classic and Modern Perfume Ingredients	178
6.7	Genetically Modified Plants	181
	6.7.1 Introduction	181
	6.7.2 Producing Genetically Modified Plants	182
	6.7.3 Impacts of Genetically Modified Plants	184
6.8	Conclusion	184
References		185

Epilogue	**188**
Glossary	**189**
How to Understand Plant Names	189
Glossary of Terms	189
Photo Credits	**197**
Subject Index	**198**

CHAPTER 1
Basic Plant Chemistry Concepts

1.1 INTRODUCTION

For millions of years plants have developed a wealth of shapes and sizes and with them a rich array of highly diverse substances that help them stay alive and reproduce. Sweet fragrances from wild roses, bright pigments in fall leaves, and potent poisons, like those of the deadly nightshade plant, are all part of the panoply of compounds that plants produce to attract, protect and repel (Figure 1.1). Most importantly, plants contain green chlorophyll, capable of trapping portions of sunlight. With chlorophyll's help, plants generate the basic chemicals that humans depend on (and could not live without), like oxygen, sugars, fats, amino acids, and vitamins.

This book is an introduction to the chemistry of plants, especially to their organic chemistry. As a preparation for the later descriptions of the chemistry of plant odors, colors, and defensive plant compounds, this first chapter reviews some basic chemistry concepts as they relate to plants.

We begin with a look at elements and their atoms. Plants need to have a set of elements available as nutrients, and in useful form. (It is a set that is not so different from human needs!) Just a couple of these elements—carbon, hydrogen, oxygen, nitrogen and a few others—assemble the wealth of carbon-based organic molecules.

The Chemistry of Plants: Perfumes, Pigments, and Poisons
Margareta Séquin
© Margareta Séquin 2012
Published by the Royal Society of Chemistry, www.rsc.org

Figure 1.1 Perfumes, pigments, and poisons.
(a) A fragrant wild rose (*Rosa rugosa*). (b) Colorful fall foliage of a grapevine (*Vitis* cultivar). (c) Branch of a deadly nightshade plant (*Atropa belladonna*). (Photo by Ruth Marent.)

Elements link up by chemical bonds to form compounds. As plants are mostly aqueous systems, a special section addresses the unusual properties of the compound water. The distinct structures of water molecules affect how water moves through plants, how minerals are transported in aqueous saps, and where pigments are stored in plant cells.

Every gardener knows that growing plants starts with the right soil. Only with the proper nutrients and conditions can plants make all the amazing compounds shown later. A discussion of soils leads to a more detailed look at ions, their mineral nutrients, and their acid or alkaline nature. Some soil compositions are also described that make plant life truly challenging.

With a basic knowledge of elements, ions, and compounds in hand, we continue to study how plant compounds interact in chemical reactions that assemble new plant compounds or break them down. Plants must be able to perform these reactions in conditions dictated by their environment, namely at ambient temperatures and mostly in water. These can be highly restrictive conditions. Yet, plants are able to function under such conditions thanks to elaborate enzymes and lots of time. Suitable nutrients, with light as the source of energy and with the help of the pigment chlorophyll, allow plants to undergo the numerous reaction steps of photosynthesis. These reactions produce oxygen and simple sugars like glucose. The sugars in turn are needed to compose all other organic compounds in plants: cellulose for plant structures, starch in bulbs to store energy, fats and amino acids, as well as plant fragrances and toxins. Respiration, the set of reactions in which sugars and fats are broken down, provide the energy for further reactions in plants.

The chapter concludes with an introduction to organic compounds and how to understand their structures. Only a few simple rules are needed to assemble basic organic molecules. Hydrocarbons, consisting of carbon and hydrogen only, will provide an introduction to organic structures. They will be illustrated with some examples of plants that contain hydrocarbons.

Just as chemical structures exactly describe the composition of a plant substance and pinpoint which compound is addressed (caffeine, vanillin, or vitamin C?), systematic names of plants, also known as scientific names or binomial names, clearly identify a plant. Common names vary regionally, and the same name may describe different plants. For example, the common name "hemlock" can refer to a poisonous, herbaceous plant or instead may describe a tree. Add to this native and foreign language names, and the confusion is complete. The scientific name of a plant, on the other hand, describes universally which plant is meant (although it can change sometimes, too, because of new studies of plant relationships). Therefore, scientific names are included with the plant examples here. The glossary at the end of this book provides explanations of key expressions (emphasized in the text) and a brief introduction to the structure of scientific plant names.

1.2 PLANTS AND THEIR ELEMENTS

For successful growth, productivity, and good survival, plants must have a continuous supply of specific nutrients which they can then transform into sugars, starch, plant structural materials, colorful pigments, and all the substances that make plant life possible. These nutrients are a collection of elements in various forms and combinations. The *periodic table* (see inside back cover) lists all the elements known to this day. But only a select few are *essential* for plant growth, meaning they must be available to plants for survival. Table 1.1 lists them and shows some of the forms or combinations in which these elements have to be available so that plants can make use of them as nutrients.[1-3] Some of their major functions in plants are shown there as well.

Plant nutrients are required in different amounts, and the quantities needed vary at different stages of plant life. Nutrient elements

Table 1.1 Essential elements in plants.

Element	Some major functions	Sources
Macronutrients		
Carbon	Essential component of organic compounds.	CO_2
Oxygen	Major component of organic compounds.	H_2O, O_2
Hydrogen	Major component of organic compounds.	H_2O
Nitrogen	Component of nucleic acids, proteins, chlorophyll, alkaloids.	NO_3^-, NH_4^+
Sulfur	Component of some amino acids, proteins, coenzymes.	SO_4^{2-}
Phosphorus	Component of nucleic acids, ATP, phospholipids, coenzymes.	$H_2PO_4^-$, HPO_4^{2-}
Potassium	For osmotic balance, operation of stomata; enzyme activator.	K^+
Calcium	Required for formation and stability of membranes; activation of some enzymes.	Ca^{2+}
Magnesium	Component of chlorophyll; activates many enzymes.	Mg^{2+}
Micronutrients		
Iron	In chlorophyll synthesis, activates some enzymes.	Fe^{3+}, Fe^{2+}
Chlorine	For ion balance; in water-splitting process of photosynthesis.	Cl^-
Boron	Cofactor in chlorophyll synthesis.	$H_2BO_3^-$
Manganese	Activates enzymes; in chlorophyll synthesis.	Mn^{2+}
Copper	Involved in redox reactions.	Cu^{2+}, Cu^+
Zinc	Activates enzymes.	Zn^{2+}
Molybdenum	Essential for nitrogen fixation.	MoO_4^{2-}
Nickel	Cofactor for enzyme in nitrogen metabolism.	Ni^{2+}

are considered to be either *macronutrients* that are required in relatively large amounts, or *micronutrients* that are needed in much smaller amounts, sometimes traces only. Nevertheless, all the nutrients must be available. The growth of plants is greatly challenged if too much of a micronutrient or too little of a macronutrient is supplied, or if toxic additional elements are present in the soil. Plants that are able to grow in such environments need to have special adaptations. Some examples are described in Box 1.1.

The *macronutrients* carbon (C), hydrogen (H), oxygen (O) and nitrogen (N) assemble most of a plant structure. They are parts of *compounds* in which the elements' *atoms* are linked to each other by *chemical bonds* to form *molecules*.[4] A large portion of hydrogen and oxygen atoms are tied up in water molecules (H_2O), as plants are mostly aqueous systems. Carbon atoms are a required part of all *organic compounds*. In addition, the elements hydrogen, oxygen, and nitrogen are the most common contributors to organic compounds. Phosphorus (P) and sulfur (S) can also be part of organic molecules. Note that the macronutrient elements mentioned are listed in vertical columns or *groups* on the right-hand side of the periodic table (with the exception of hydrogen). These are the groups that contain *non-metal elements*. Their atoms bond to form the organic molecules of plant structures, of plant odors, of pigments, and of defensive substances in plants. In this book we will mostly encounter organic compounds.

Two gases dominate Earth's atmosphere: nitrogen gas (N_2), contributing about 78% of it, and oxygen gas (O_2), providing about 21% of our atmosphere. Oxygen molecules are produced by plants during photosynthesis. They participate in the breakdown of plant molecules during respiration, providing energy for plant processes. Therefore, oxygen has vital roles not only as part of compounds in plant life, but also as oxygen gas.

One might assume that nitrogen gas, as the largest component of air, would be a major direct supplier of the macronutrient nitrogen. Yet, plants cannot make direct use from air of this much-needed nutrient. Nitrogen molecules are unreactive, because of a strong triple bond between the nitrogen atoms (Figure 1.10(d)). Therefore, plants need the help of beneficial bacteria that convert nitrogen gas into forms that plants can work with, like ammonium (NH_4^+) and nitrate (NO_3^-) ions. Many different plants have root nodules with nitrogen-fixing bacteria, as shown in Figure 1.2(a). Plants in the legume family, like peas and beans, have such root nodules and are

(a) (b) (c)

Figure 1.2 Special plant structures to obtain useful nitrogen.
(a) Root nodule in the roots of *Pultanea pedunculata*, an Australian legume. (The red arrow points to the nodule.) (b) and (c) Insectivorous plants: (b) Venus Flytrap (*Dionaea muscipula*) and (c) Cobra Lily (*Darlingtonia californica*).

well-known for their nitrogen-fixing abilities. Farmers of former times carefully incorporated plantings of them in their crop rotations, to get the nitrogen into their soils. Some plants have even learnt to trap and digest insects to obtain yet another source of useful nitrogen. Insectivorous plants, in their amazing trapping mechanisms, have special enzymes that digest and break down the ensnared insects and their nitrogen-containing proteins and thus obtain additional useful nitrogen compounds (Figures 1.2(b) and 1.2(c)).

On the left-hand side and towards the bottom of the periodic table we find elements that are *metals*. Among them we find the macronutrients potassium (K), calcium (Ca), and magnesium (Mg). They mostly function as *ions*, *i.e.* as atoms with a charge, in the plant system. Metal ions have positive electrical charges; they are *cations*. Elements that are in the same group of the periodic table tend to have similar properties. Calcium and magnesium, for instance, both in main group number two, form ions with a 2^+ charge and react similarly. Elements of the same group therefore can sometimes take each other's place in compounds or reactions, as shown in the plant examples in Box 1.1.

> **BOX 1.1 Plants that can Deal with Unusual Elements**
>
> While all plants need the complete set of required elements, there are some specialist plants that can pick up and deal with otherwise harmful elements. The stinking milk vetch (*Astragalus praelongus*), from North America's Southwest, is an example (Figure 1.3(a)). This plant is capable of absorbing toxic selenium (Se) from the soil, by replacing sulfur (S) in some of its amino acids.[5] Note that the elements selenium and sulfur are in the same group in the periodic table and therefore have similar properties. Stinking milk-vetch plants are able to further process the selenium compounds and get rid of them as hydrogen selenide gas (H_2Se). This gas is just as malodorous as hydrogen sulfide (H_2S), with the smell of rotten eggs.
>
> Some other specialist plants can tolerate unusually high concentrations of metals in soils. A few types of plants are even capable of surviving on mine tailings and are used to revegetate and stabilize soils in these environments.
>
> Serpentine areas, with rocks that often have shiny surfaces and sometimes a greenish appearance, are found in specific areas

Figure 1.3 Plants adapted to unusual elements in soils. (a) Stinking milk-vetch (*Astragalus praelongus*), adapted to selenium salts in the soil. (b) A serpentine-tolerant wild onion (*Allium falcifolium*) growing among serpentine rocks.

all over the world (Figure 1.3(b)).[6] Their soils have an unusually high content of magnesium compared to calcium. (Note that magnesium and calcium are in the same group of elements in the periodic table.) In addition, these soils tend to have high amounts of heavy metals, like iron, copper, or nickel, that are toxic to most plants. Yet, again, some specialist plants, such as the wild onion shown in Figure 1.3(b), have adapted to these environments and have found a biological niche to thrive in.

Only traces are needed of the micronutrients. Among them we find iron (Fe), copper (Cu), manganese (Mn), zinc (Zn) and molybdenum (Mo). They are metallic elements listed in the short vertical groups in the center of the periodic table, known as transition elements. Two additional essential micronutrients are chlorine (Cl) and boron (B). Chlorine, a nonmetal, must be supplied as chloride (Cl^-) ions. The name for negatively charged ions is *anions*. Boron is usually part of more complex anions. (See Table 1.1.)

The soils that plants grow in supply most of the plant nutrients. The exception is the element carbon which plants obtain from air as carbon dioxide (CO_2). Less than one percent of air is carbon dioxide, and it is the source of carbon in photosynthesis. A later section in this chapter will discuss different types of soils and their properties.

Decomposing plant materials are great suppliers of essential elements in useful form. In tropical rain forests they provide the major source of plant nutrition. If rain forest is logged and burned down, there is initially a good nutrient supply from all the former plant materials. But this rich supply of mineral nutrients is rapidly depleted, as it is washed out by rains and consumed by plants, with no chance for renewal.

Plant nutrients can be made available by wildfires as the heat breaks down complex plant compounds into simpler ones that become part of the soil. If wildfires get too hot, though, nutrients like nitrogen are lost as they escape as gaseous nitrous oxides into the air. Some natural environments that regularly experience wildfires have their collection of plants that are fire followers. Plants like fireweed (*Epilobium* spp.) take advantage of the increase in sunlight and all the newly available nutrients from the ashes after an area has burned (Figure 1.4(a)).

If soils are lacking adequate amounts of some of the macro- or micronutrients, farmers and gardeners tend to apply fertilizers. Commercial fertilizer packages characteristically display three numbers on their labels, like "10-4-10" or "4-12-4", which refer to three of the most needed macronutrients, N (nitrogen), P (phosphorus) and K (potassium) (Figure 1.4(b)). Fertilizers must contain them in the form of compounds or ions that plants can make use of, and the numbers relate to the relative weight percent of such compounds. If soils do not supply the essential nutrients in adequate amounts, plants will turn yellow, shrivel up, have irregular growth, or will die. Too much of a nutrient can also kill a plant,

Figure 1.4 Plants and nutrient elements.
(a) Pink fireweeds (*Epilobium angustifolium*), found throughout the temperate northern hemisphere, take advantage of nutrients made available after a forest fire. (b) Fertilizer packages usually indicate their N-P-K content by a set of three numbers.

as some unlucky gardeners experience after applying too much nitrogen-rich fertilizer.

It is interesting to compare which of the essential plant elements are also required by the human body. Except for boron, humans need all of the plant elements as well. But in addition, the human diet must provide traces of fluorine (F) and iodine (I), both as anions. Humans also need additional mineral ions, supplied in the form of cations: they are sodium (Na), and traces of cobalt (Co), chromium (Cr), and selenium (Se).

1.2.1 Elements and Atoms

This section serves as a brief review of the structure of atoms as their structures determine how atoms combine by chemical bonds and form molecules. Elements are composed of *atoms*, the units that compose all matter. Atoms are extremely small; it would take about a million of them to stretch across the period printed at the end of this sentence.[7] Nevertheless, their structures determine which element they belong to (hydrogen or carbon or oxygen?), or which elements can form bonds with each other and link up to compounds. The periodic table provides a lot of information about atomic structures.

At the center of each atom is a tiny, highly condensed nucleus that accounts for almost all of an atom's mass. In this nucleus are the positively charged *protons* and neutral particles called *neutrons*, as shown in the simple models of a helium (He) atom and a carbon atom in Figure 1.5. They are the subatomic particles that determine the mass of an atom, the atomic mass unit for a proton or for a neutron being 1 amu or 1. The number of protons in the nucleus of an atom determines the identity of an element. Therefore, an element whose atoms have one proton only in its nuclei is always hydrogen. The atoms of the element helium are characterized by two nuclear protons and those of carbon atoms by six protons (Figure 1.5). The number of protons in the atoms of a specific element can be read from the periodic table: it is the same as the *atomic number* of an element. It also tells us the position of an element in the chart and the number of *electrons* around the nucleus in an uncharged atom.

While all atoms of an element have the same number of protons, the number of neutrons in their nuclei may vary. *Isotopes* are atoms

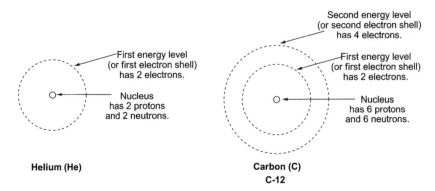

Figure 1.5 Simple models of a helium and a carbon atom.
A helium (He) atom has two protons and two neutrons in its nucleus and two electrons in its first energy level or electron shell. A carbon atom has six protons in its nucleus. The most common isotope of carbon, C-12, has six neutrons. A neutral carbon atom has a total of six electrons in two electron shells or energy levels: two electrons in the first shell, and four in the second shell. (The models are not to scale.)

of the same element, but with different numbers of neutrons. This means their masses are different too. Some isotopes are stable and very common, like carbon atoms with six neutrons in their nuclei. Together with the masses of their six protons they have a mass of 12 atomic mass units, written as carbon-12, C-12, or ^{12}C (Figure 1.5). Carbon-14 or ^{14}C is another isotope of carbon. Its atoms have eight neutrons and therefore a mass of 14. Atoms of carbon-14 are unstable and decay spontaneously; they are *radioactive*. ^{14}C is used in radiocarbon dating, and also in studies that research how plants assemble organic compounds in plant reactions. Oxygen atoms have eight protons and most commonly eight neutrons in their nuclei. A rare, heavier isotope, ^{18}O, has ten neutrons and is also stable. We will encounter it in studies on photosynthesis.

In a neutral atom, the total number of negatively charged *electrons* is equal to the number of protons. As the mass of an electron is much smaller than the mass of a proton or neutron, electron masses are usually ignored when calculating the total mass of an atom (as was done for determining the masses of carbon isotopes in the previous paragraph).

Electrons are moving at specific distances around the nucleus, with fixed levels of potential energy, called *energy levels*, or *electron shells* (Figures 1.5 and 1.6(a)). The more energy electrons have, the

Basic Plant Chemistry Concepts

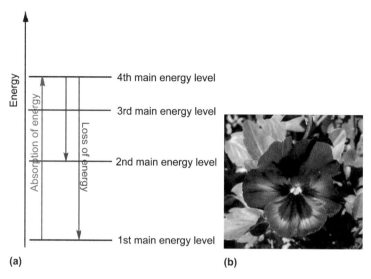

Figure 1.6 Energy levels of electrons.
(a) Electrons in an atom are arranged in defined energy levels, or electron shells, around the nucleus. If electrons absorb energy in the form of light, they are excited to higher energy levels (shown with red arrow). They will drop to lower energy levels again with release of energy (shown in blue). (b) Plant pigments, as in the leaves and flower of a viola, have molecules whose electrons are easily excited to higher energy levels and in the process absorb light.

further away they are from the atom's nucleus, and the less they are attracted to the positive charges of the protons. Energy, *e.g.* in the form of light, can excite electrons and move them to higher energy levels, a process that involves absorption of energy. (When electrons drop again to lower original energy levels, energy is released, usually in the form of very small amounts of heat.) Colorful plant pigments, such as in Figure 1.6(b), are composed of molecules with many excitable electrons in their bonds. These electrons are easily lifted to higher energy levels and absorb light energy in the process. The later chapter on plant pigments will introduce typical bonding patterns in molecules of colorful compounds.

As a preparation for chemical bonding between atoms, let us look at electrons in more detail. The periodic table shows how many energy levels of electrons are found in the unexcited or *ground state* of an atom. With each start of a horizontal row of elements in the table, called a *period*, a new shell or main energy level is added. The first electron shell can accommodate a maximum of two electrons, the second one a total of eight. Electrons in the outermost shells, called the *valence shells*, accomplish most of the bonding between atoms as they are furthest away from the attractive forces of the nucleus. The group number of an element in the periodic table's main groups is equal to the number of *valence electrons*. Carbon is in main group number four, and therefore carbon atoms have four valence electrons (Figure 1.5).

Within their main energy levels, electrons occupy *orbitals*. The latter can be described as the three-dimensional spaces where an electron can be found with about 90% probability (Figure 1.7). Each orbital can hold a maximum of two electrons. The first two electrons of each energy level are in an orbital that is spherical in shape; it is called the *s-orbital* (Figure 1.7(a)). Hydrogen and helium atoms have their electrons in *s*-orbitals. Starting with the second electron shell, the first two electrons of this energy level are again in an *s*-orbital, and the next six electrons occupy three barbell shaped orbitals called *p-orbitals*, oriented along the *x*, *y*, *z* axes in space (Figures 1.7(b), (c)).

As this book is mostly concerned with organic compounds, we focus on the orbitals of a carbon atom when carbon bonds with four other atoms (Figure 1.7(d)). The four valence electrons of carbon are located singly in four equivalent orbitals. These orbitals are combinations, or hybrids, between *s*- and *p*-orbitals. As they all contain negatively charged electrons, they repel each other, and, as a consequence, are as far apart from each other as possible, which

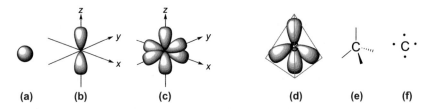

(a)   (b)   (c)   (d)   (e)   (f)

Figure 1.7 Electron orbitals and carbon.
(a) The spherical shape of an *s*-orbital. (b) The barbell shape of a *p*-orbital. (c) Three *p*-orbitals oriented along the *x*, *y*, *z* axes. (d) The tetrahedral arrangement of the four hybrid orbitals of carbon when carbon bonds to four other atoms. (A tetrahedron is shown in red.) (e) A wedge-and-dash picture of the tetrahedral arrangement of carbon bonds. (f) An electron-dot picture of carbon with its four valence electrons.

Basic Plant Chemistry Concepts

is towards the corners of an imaginary tetrahedron (shown in red in Figure 1.7(d)). This orbital arrangement determines the shape of molecules in which carbon atoms form single bonds to four other atoms. The molecular shapes in turn affect how molecules interact with each other. Figure 1.7(e) shows a "wedge-and-dash" picture that is commonly used to point out the tetrahedral arrangement of bonds around a carbon atom in a molecule. An even simpler picture of carbon with its four valence electrons is the electron-dot structure shown in 1.7(f); it can be used to explain the bonding of carbon, as shown in the next section.

1.2.2 Chemical Bonding to Form Molecules

Electrons, namely the valence electrons, do the bonding between atoms. When two atoms form a chemical bond, they either lose or gain electrons from their valence shells and become ions, or they share electrons and form *covalent bonds*. By forming a chemical bond, atoms strive to obtain a stable set of electrons with a full valence shell, *i.e.* a *noble gas configuration*. By losing or gaining electrons, or by sharing electrons, atoms each obtain eight electrons in their valence shells, with the exception of hydrogen which can have a maximum of two electrons.

Many nutrients in soils are supplied as *ions*. When atoms of metal elements form cations, they lose electrons from their valence shells. This results in an imbalance of protons and electrons and creates atoms with an overall positive charge. Examples are ions of sodium (Na^+), potassium (K^+), calcium (Ca^{2+}) or aluminum (Al^{3+}). Atoms of some nonmetal elements, like chlorine, acquire additional electrons to reach a full valence shell. This results in a surplus of negative charges and the formation of negatively charged anions, like Cl^-. In soils, nutrients to plants can also be in the form of complex ions, like ammonium (NH_4^+) ions or anions like phosphates (PO_4^{3-}) or carbonates (CO_3^{2-}). Cations and anions attract each other due to electrostatic forces and form *ionic bonds*. They combine in fixed proportions to form *ionic compounds*, also known as *salts*. Ions in solid salts form three-dimensional, regular arrangements, called lattices. Figure 1.8 illustrates the formation of an ionic bond between a sodium and a chlorine atom and shows part of the lattice pattern between sodium and chloride ions. In formulas of ionic compounds, like $NaCl$, $CaCO_3$, or K_3PO_4, the overall negative and positive charges of the ions are balanced, and number subscripts show the correct proportions.

Atoms of non-metal elements, like carbon, hydrogen, or oxygen, form bonds with other non-metal atoms by sharing electrons from their valence shells. Their orbitals overlap and combine to form bonding orbitals. This type of bond is called a *covalent bond*. Electron-dot structures can be used to illustrate the shared valence electrons. Figure 1.9 shows structures of hydrogen (H_2) and methane (CH_4) molecules. Each molecule is shown as an electron-dot structure, with its shared valence electrons doing the bonding, and as a "ball-and-stick" model that shows the shape or geometry of the molecule. For methane, a "wedge-and-dash" picture is shown as well. We will find covalent bonds almost exclusively in the molecules of this book: in water molecules, and in all the molecules of sugars, fats, plant odors, pigments, and toxins.

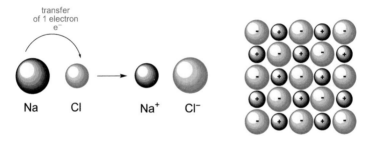

Figure 1.8 Ionic bonds.
The transfer of an electron from a sodium atom (Na) to a chlorine atom (Cl) forms a sodium ion (Na^+) and a chloride ion (Cl^-), forming sodium chloride. Cations and anions are arranged in regular, three-dimensional lattices in salts.

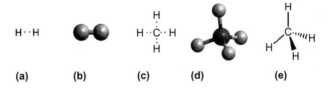

Figure 1.9 Covalent bonds.
A hydrogen molecule (H_2) is shown as (a) an electron-dot structure, and (b) as a "ball-and-stick" model. In (c), a methane molecule (CH_4) is presented as an electron-dot structure. The tetrahedral arrangement of bonds in methane is pointed out in the "ball-and-stick" model in (d) and in the "wedge-and-dash" picture in (e).

Basic Plant Chemistry Concepts

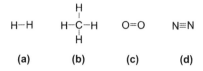

Figure 1.10 Molecules with covalent bonds. Molecules of (a) hydrogen (H_2) and (b) methane (CH_4) have single bonds. A double bond is shown in (c) oxygen (O_2), and a triple bond in (d) nitrogen (N_2).

Single covalent bonds, composed of two bonding electrons, are most easily drawn as a single line between atoms. Figure 1.10 shows single bonds in the molecules of hydrogen (H_2) and methane (CH_4). Double bonds, with four electrons bonding, are written as double lines, as in the molecule of oxygen (O_2). Triple bonds are written as triple lines, as shown in the structure of a nitrogen molecule (N_2).

1.3 LIFE-GIVING WATER

Every living plant and animal on earth needs water. It is the most abundant liquid on earth, and its properties affect all living things. Because water is so common, we often overlook how unusual a compound it is.

Here are a couple of its unusual properties:

- Water is a liquid over a very large temperature range, from 100 °C (212 °F) all the way down to 0 °C (32 °F). It is water in its liquid form that is useful to plants.
- Water heats up—and cools down—very slowly compared to other liquids. This is an important advantage to plants that live in or near water and that cannot move away when rapid temperature changes occur.
- Water has a tendency to cling to surfaces. It has a high surface tension, higher than most other liquids. Water molecules cling to each other, too. This allows plants to transport water from their root hairs to the tips of their leaves.
- Liquid water has a higher density than its solid form, a very unusual property among compounds. Because of this, ice will float on water, and ponds and lakes freeze from the top down, not the other way round.

A closer examination of water molecules provides explanations for these unique properties. H_2O molecules have an angular shape (Figure 1.11). Hydrogen and oxygen atoms form a covalent bond by sharing electrons, the bonding orbitals each obtaining a pair of electrons. But the shared electrons are not equally distributed between the atoms. It is the nature of the compact, *electronegative* oxygen atom to pull the bonding electrons somewhat closer, away from the hydrogen atoms. This gives oxygen atoms in water molecules a partial negative charge (δ^-), leaving a slightly positive charge on the hydrogen atoms (δ^+). As a result, water molecules are *polar*, and the bonds in them are *polar covalent*.

When water molecules approach each other, the slightly positive ends become attracted to the oxygen atoms, and the water molecules become connected to each other in a huge network. The bonds between water molecules are called *hydrogen bonds* (Figure 1.12). Their strength is much weaker than covalent bonds or bonds between ions. Yet, hydrogen bonds determine many properties of

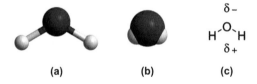

Figure 1.11 Water, an extraordinary compound.
A water molecule, shown (a) as a ball-and-stick model, (b) as a space-filling model, and (c) as a line structure, has an angular structure. The partial negative charge (δ^-) on oxygen and the partial positive charges (δ^+) on the hydrogen atoms make water molecules polar.

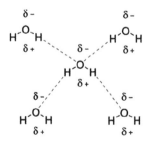

Figure 1.12 Hydrogen bonding of water molecules.
Polar water molecules are attracted to each other and connect by hydrogen bonds (shown in red as broken lines).

water. They are also most important bonds within very large molecules, like proteins or nucleic acids, where they determine the shapes of the molecules and with this their functioning.

Hydrogen bonding explains many of the extraordinary phenomena relating to water. The huge network of interconnected molecules in liquid water requires large amounts of energy to absorb heat, much more so than separate molecules would. Therefore, a body of water, like a lake, heats up and cools down slowly. At 4 °C, water has its highest density. When it cools further down and forms ice, there is still hydrogen bonding between the H_2O molecules; but the molecules are arranged more loosely. This results in a lower density of ice than of water, and ice floats on water.

In plants, water is picked up through the root hairs (Figures 1.13 and 1.14(a)), and the plants seem to "pull up" the water. Actually, as water molecules cling strongly to each other through hydrogen bonding, they move in an uninterrupted network, due to capillary action, from the root hairs all the way to the leaves where large amounts of water evaporate through tiny openings called *stomata* (singular: *stoma*) (Figures 1.13 and 1.14(b)).[8]

Water is often called the *universal solvent* because it can dissolve more substances than any other liquid. The polar nature of water explains why it can dissolve salts and sugars and other polar compounds. The transport of compounds in the aqueous plant system is determined by their water-solubility (or insolubility). Water can dissolve many salts from the soil. When an ionic compound, like sodium chloride, is placed in water, the polar water molecules are attracted to the individual ions (Figure 1.15). The partially negative charges on oxygen atoms in water molecules are attracted to cations, like sodium ions (Na^+), whereas the partially positive charges on the hydrogen atoms of water are attracted to anions, like chloride (Cl^-), as shown in Figure 1.15. Water molecules thus surrounding ions in a solid ionic compound break up the lattice and dissolve the salt. Plants can easily absorb nutrients in the form of aqueous solutions. Ionic nutrients dissolved in water can be pulled up by the root hairs and follow the route of water molecules through a system of vascular tissues called the *xylem* (Figure 1.13).

The old chemistry saying "like dissolves like" helps us to remember that polar water can dissolve other polar compounds. Sugars, like glucose, are polar organic compounds. Glucose and

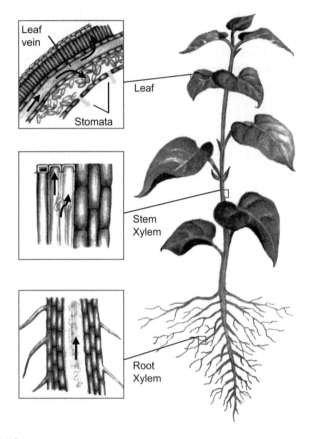

Figure 1.13 Water movement in a plant.
(Drawings by Eveline Larrucea.)

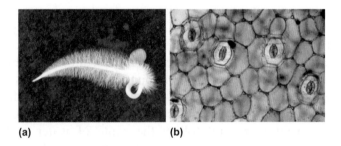

Figure 1.14 Root hairs and stomata.
(a) Root hairs of a radish seedling. (b) Stomata from a leaf of *Rhoeo* sp. (Light micrograph © Institute of Botany, University of Innsbruck, Austria.)

Basic Plant Chemistry Concepts 19

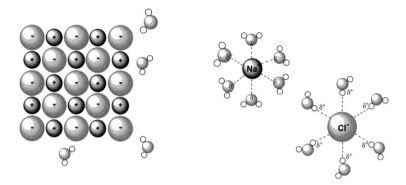

Figure 1.15 Water molecules surrounding ions.
The partially negative charges on the oxygen atoms of water molecules surround sodium ions (Na^+). The partially positive charges on the hydrogen atoms are attracted to chloride ions (Cl^-).

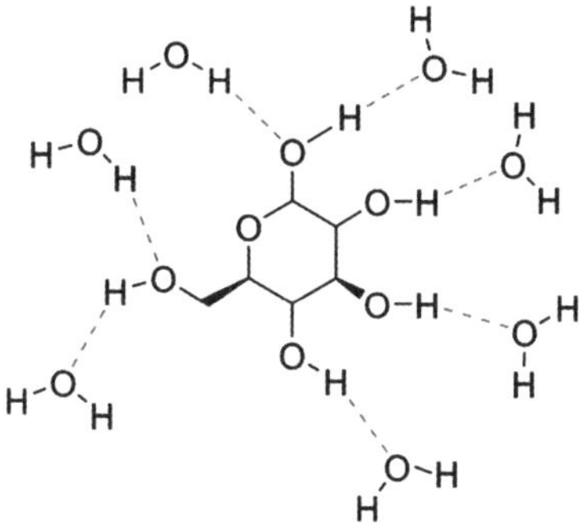

Figure 1.16 Glucose in water.
Water molecules form hydrogen bonds (shown in red) with the OH groups of a glucose molecule.

other sugars have numerous OH groups that water molecules can form hydrogen bonds with, as shown in Figure 1.16. Surrounded by water molecules, the sugar molecules become separated from the solid crystals and dissolve. Sugars and other water-soluble organic molecules can be transported in plants as aqueous solutions, in vascular tissues called the *phloem*.

Water is a poor solvent for non-polar substances like oxygen (O_2). If molecules have an even distribution of bonding electrons as in O_2, polar water cannot surround them by forming hydrogen bonds. Yet, plants (and animals) that live in water need adequate

(a) (b) (c)

Figure 1.17 Water plants and oxygen.
(a) The large leaves of water lilies (*Nymphaea* cultivar) keep water cool, allowing for more oxygen to dissolve in water. (b) Air chambers, shown in sliced stems of tule reeds (*Scirpus acutus*), allow for uptake of air oxygen. (c) *Elodea* water plants produce oxygen in water.

amounts of oxygen for vital metabolic processes like respiration. To obtain the needed oxygen, plants that live partially in water have special adaptations. Water lilies (Figure 1.17(a)) have large, waxy leaves that shade the water, keeping it cool and allowing some air oxygen to dissolve. Reeds, like tules (*Scirpus acutus*, Figure 1.17(b)), have stems filled with spongy air chambers that can transport air oxygen to water-locked plant parts. (These structures made tules useful materials for building boats as the reeds float on water.) Plants that are completely immersed in water, like *Elodea* (Figure 1.17(c)), depend on an initial supply of oxygen in water. Once growing, these plants produce an even larger amount of oxygen through photosynthesis, and water plants like *Elodea* are important suppliers of oxygen for aquaria and ponds.

In pure water, about one out of every 550 million water molecules (this means very few) separate or *dissociate* into hydrogen ions (H^+) and hydroxide ions (OH^-) (Figure 1.18 and Equation (1.1)). The ions are surrounded by water molecules, shown by appending *(aq)*. Between the ions and H_2O molecules there is a *dynamic equilibrium*, meaning there is a continuous back-and-forth movement between the separated ions and the water molecules, with most of the reaction on the side of the undissociated H_2O (indicated by the direction of the bold arrow).

As all life on earth is based on water, the balance or imbalance of hydrogen and hydroxide ions influences life in general and plants in

Basic Plant Chemistry Concepts

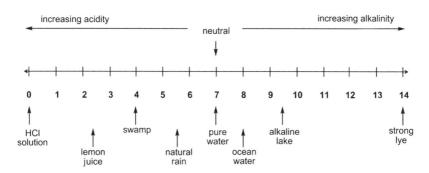

Figure 1.18 Dissociation of water molecules.
Only about one in 550 million water molecules dissociates into a hydrogen ion and a hydroxide ion.

Figure 1.19 The pH scale.

particular. If an aqueous system has a surplus of hydrogen ions, it is called *acidic*, and *acids* are compounds that release hydrogen ions when dissolved in water. In contrast, *bases* release hydroxide ions into water, and a high concentration of hydroxide ions leads to an *alkaline* or *basic* medium. The measure of the *pH* is based on water. In everyday life, we hear of acid rain, or of the pH of foods, or of plant soils needing the proper pH. The *pH-scale* (Figure 1.19) is a measure of acidity or alkalinity. The scale ranges from 0–14, with pH 7 standing for neutral and representing the pH of pure water and a balance of hydrogen and hydroxide ions. A pH value below 7 refers to an acidic medium. Above 7 means alkaline or basic. The pH scale is a logarithmic scale, with a factor of ten between each pH unit. Therefore, a medium that has a pH of 6 is ten times more acidic than one with a pH of 7, and water in a swamp or bog with a pH of 4 is a thousand times more acidic than pure water.

Natural rain, without any pollutants, has a pH around 5.6, because CO_2 from air mixes with water to form carbonic acid, H_2CO_3 (Figure 1.20 and Equation (1.2)). As a *weak acid*, carbonic acid is only partially dissociated into ions (note the back reaction to the undissociated H_2CO_3).

$$H_2O + CO_2 \rightleftharpoons H_2CO_{3\,(aq)}$$
$$H_2CO_{3\,(aq)} \rightleftharpoons HCO_{3\,(aq)}^- + H^+_{(aq)}$$
(1.2)

Figure 1.20 Natural rain.
Carbon dioxide from air mixes with rainwater, forming carbonic acid. This makes natural rainwater slightly acidic. Note the formation of hydrogen (H^+) ions.

$$CaCO_3 \rightleftharpoons Ca^{2+}_{(aq)} + CO_{3\,(aq)}^{2-}$$
$$CO_{3\,(aq)}^{2-} + H_2O \rightleftharpoons HCO_{3\,(aq)}^- + OH^-_{(aq)}$$
(1.3)

Figure 1.21 Calcium carbonate dissolving in water.
When calcium carbonate ($CaCO_3$) dissolves in water, it first dissociates into ions. The carbonate ions (CO_3^{2-}) react further with water, leading to the formation of hydroxide (OH^-) ions and an alkaline medium.

In the desert, lakebeds without any outlets tend to be strongly alkaline. As water evaporates, high concentrations of salts like calcium carbonate accumulate in the lake water. The dissolving salts react further with water and produce hydroxide ions, as shown in Equation (1.3) in Figure 1.21.

Plants themselves contain many different organic acids and bases. We notice the sour taste of acids like citric acid in lemons, or oxalic acid and malic acid in rhubarbs (Figure 1.22(a)). These acids release hydrogen ions into the aqueous plant saps. As they are weak acids, they do not affect the pH of the plant dramatically. Strong acids that are fully dissociated into ions, like hydrochloric acid, are not found in plants. As for bases in plants, the most famous ones are likely to be the *alkaloids*, like caffeine in coffee plants (Figure 1.22(b)). Alkaloids are organic plant compounds that often act as defensive substances in plants. They have nitrogen atoms in their structures and tend to be weakly basic when dissolved in the aqueous plant saps. We will encounter them in the chapter on plant poisons.

Water and its properties affect all aspects of plant life. Last but not least: water is required for photosynthesis. The splitting of

Basic Plant Chemistry Concepts 23

(a)　　　　　　　　　　(b)

Figure 1.22 Plants with organic acids or bases.
(a) Rhubarb stalks and leaves (*Rheum rhabarbarum*) contain oxalic acid and malic acid, both organic acids. (b) Coffee plants (*Coffea* sp.) contain caffeine, an alkaloid plant base.

water molecules, accomplished by green plants, algae, and some bacteria during photosynthesis, provides living things with oxygen and with the basic sugars needed for composing further vital compounds. Water is truly a life-giving compound.

1.4 NUTRIENTS FROM THE SOIL

The soils that plants grow in supply water and mineral nutrients. Soils are highly diverse, complex mixtures. They contain tiny rock fragments composed of ionic compounds, dead and decaying plant and animal materials, and diverse societies of organisms (Figure 1.23).[9] When moisture in the soils dissolves some of the salts and rocky materials, plants can pick up nutrients in solution through their root hairs. The structure of the soils that plants grow in, their mineral content, water-holding ability, aeration, and their acidity or alkalinity all determine how well a plant will grow in them.

Sandy soils (Figure 1.23(a)), with relatively large particles of quartz (which is silicon dioxide, SiO_2), have equally large air spaces between the particles. This allows water to pass freely between the sand grains and to run off rapidly. The result is that little water is retained, and dissolved nutrients drain off fast, too. As a benefit to plants, sandy soils are well aerated and promote the processes of respiration and of healthy growth of plant roots. In contrast, the soil type that we know as clay (Figure 1.23(c)) consists of very fine particles, called *micelles*. Negatively charged ions, like silicate (SiO_3^{2-}), aggregate at the surface of the micelles and attract cations like potassium ions (K^+),

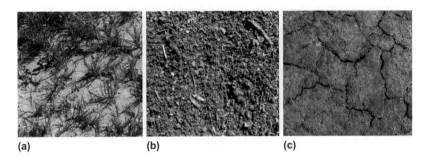

(a) (b) (c)

Figure 1.23 Different types of soils.
(a) Sandy soil with dune grasses. (b) Close-up of loose compost, with decomposing plant materials. (c) Cracked clay soil.

magnesium ions (Mg^{2+}), and calcium ions (Ca^{2+}). Plants in clays get adequate supplies of these nutrients. But the charged surfaces of clay micelles hold tenaciously to polar water molecules. For that reason, clays turn soggy in heavy rains, leaving root systems with little oxygen. When clays dry out, they form cracked surfaces. The experienced gardener adds good compost (Figure 1.23(b)) to clay soils, with lots of decomposing plant material, to loosen up the soil and make water and minerals better available to plants. Compost is rich in humic acids, the deep-brown to black biomaterials in fertile soils.[10,11] (Their name is derived from "humus" which is Latin for "earth".) Humic acids are products of degradation of plant and animal materials and are composed of complex organic macromolecules. Humic acids help regulate soil pH, retain water in the soil, bind metal ions, and can transform and absorb toxic pollutants. They are thus highly desirable materials in soils.

Plants are very sensitive to the pH of the soils they grow in. Many plant processes involve hydrogen ions, and soil pH affects them. The environment around roots is slightly acidic as the fine root hairs give off CO_2 as a product of respiration. According to Equation (1.2), soil moisture combines with CO_2 to form carbonic acid. The released hydrogen ions can be exchanged for metal ions. Most plants do best in soils with a pH not too far from neutrality, with some plants, like azaleas and rhododendrons, needing acidic soils, with a pH between 4.5 and 5.5, and some vegetables preferring a somewhat alkaline soil. If the soil pH is high, essential metal ions, like iron ions, get tied up as water-insoluble hydroxides and are not available to plants anymore. On the other hand, acidic

(a) (b)

Figure 1.24 Environments with different soil pH.
(a) Acidic soils are found in forests with ample rainfall, as in a temperate rain forest. (Photo by Glenn Keator.) (b) Alkaline soils are found in limestone and dolomite formations, as in the dolomite area of the White Mountains, California.

soils make metal ions well available. A very low pH frees too high a concentration of metal ions for healthy plant growth.

Environments with ample rainfall, like a rain forest (Figure 1.24(a)), tend to have acidic soils because mineral salts that can neutralize the slightly acidic rainwater are continuously washed out. Limestone and dolomite areas (Figure 1.24(b)), on the other hand, have alkaline soils. Their carbonates slowly dissolve in water (it is in limestone areas that we find caves), and in the process form hydroxide ions in water (Equation (1.3) in Figure 1.21).

Strongly acidic soils that have high metal concentrations can be found close to some springs (Figure 1.25(a)) or in mine tailings. These environments are either devoid of plants, or only specialist plants can survive in them. High soil alkalinity, with pH values of 9 or more, is found in arid areas, like alkaline lakebeds (Figure 1.25(b)) of deserts, where rains are scarce and salts builds up.

While some salts, like carbonates, affect the pH when they dissolve, others, like sodium chloride, dissociate into ions without further reactions with water and do not influence the pH. Nevertheless, the high concentrations of ions in soils destroy the *osmotic balance* that must exist between plant cells and their aqueous surroundings. The membranes and cell walls that bound living plant cells (Figure 1.26) are not solid barriers. Water molecules diffuse freely back and forth through them. As if to even out differences in salt content on the inside of the cell and the outside, water molecules tend to diffuse to the side where there is the higher salt concentration, in a process called *osmosis*. In salt-water marshes and in

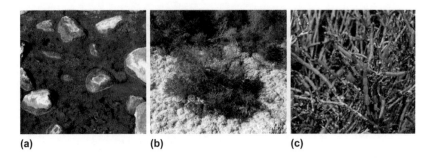

Figure 1.25 Plants and extreme soils.
(a) Acids from a spring, caused by carbon dioxide bubbling up, dissolve high concentrations of iron. This results in an area devoid of plants. (b) Bush pickleweed (*Allenrolfea occidentalis*) is a desert plant that is highly adapted to strongly alkaline and salty lakebeds. (c) Pickleweed (*Salicornia* sp.) is a common salt-tolerant plant in salt marshes.

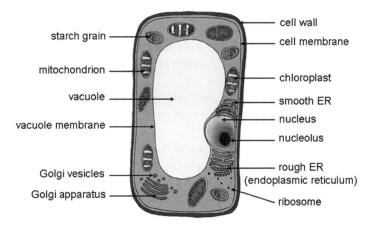

Figure 1.26 A plant cell.
(Drawing by Eveline Larrucea.)

salty soils of the desert, we find a higher salt concentration on the outside of plant cells that bound the salty water, and water molecules have a great tendency to move out of the plant cells. This makes the cells more and more desiccated and shriveled up. Only plants with highly specialized cell mechanisms can survive in such environments, and plant names like pickleweed or *Salicornia* (*sal* being Latin for salt) reflect these adaptations (Figure 1.25(c)).

Sodium chloride and similar salts in soils are a serious problem in agriculture. Irrigation leaches salts out of the soils and into the groundwater. Repeated re-use of water leads to increasing soil salinity and makes the soils useless for growing crops.

1.5 PLANTS AND CHEMICAL REACTIONS

Many subtle and complex chemical reactions in plants change the rate of growth of stems and leaves, the color of their fruits, the intensity of odors in blossoms, or the decomposition of their leaves. During photosynthesis plants trap carbon dioxide and, with water, light and chlorophyll, transform it into glucose. These plant processes involve chemical reaction steps in which compounds interact with each other to form new compounds.

Whenever a *chemical reaction* occurs, whether it is in a laboratory or in a living organism like a plant, bonds between atoms get broken and new bonds are formed, leading to new combinations of atoms and new molecules. The chemical compositions of the materials at the beginning of the reaction, the *reactants*, and of the resulting *products* are different. This is expressed in their different chemical formulas. *Energy* is required to get the reactions going, mostly in the form of heat energy. Other forms of energy, like light, can induce reactions, too. Some chemical reactions release energy overall; they are *exergonic* reactions (*exothermic* if the release is in the form of heat energy). Other chemical processes need a steady input of energy to keep going and are called *endergonic* (or *endothermic*) reaction.

To apply these statements to chemical processes occurring in plants let us consider the general equation for photosynthesis (Equation (1.4)).

$$6CO_2 + 6H_2O \longrightarrow C_6H_{12}O_6 + 6O_2 \qquad (1.4)$$

The composition of the reactants on the left side of the equation and of the products on the right side is obviously very different. Sunlight supplies the energy to get the process started and is required to keep photosynthesis going.

Chemical reactions in plants encounter restricted conditions: the reactions have to occur at ambient temperatures given by the environment, and many plant reactions have to take place in *aqueous* plant saps, with the reactants dissolved in water. (In contrast, a chemical reaction that is carried out in a laboratory can be performed in any suitable solvent and can be accelerated by heating the mixture.) In spite of restrictions, plants produce most elaborate products. How do plants accomplish this? With time and highly specialized catalysts! Plant reactions are complex multistep reactions that progress in controlled reaction steps.

Catalysts are chemical compounds that speed up chemical reactions by lowering the activation energy, *i.e.* the energy that is required to get the reaction going. Catalysts interact with a reactant and form an activated complex with it that in turn will react much faster with other reactants. The catalysts then separate or regenerate to interact with more reactants. Therefore, catalysts are not used up during a chemical reaction. Catalysts in living organisms are called *enzymes*. They are among the most sophisticated substances. They are often large protein molecules with defined three-dimensional shapes, which enable them to interact with only certain reactants, and only in specific ways or orientations.

Chemical reactions in living organisms that are catalyzed by enzymes are known as *metabolism*. Metabolic reactions can involve the synthesis of larger molecules (anabolism) or the breakdown of larger molecules into smaller ones (catabolism).

There are many types of chemical reactions. The processes of photosynthesis and of respiration involve numerous *oxidation-reduction* reactions. The *oxidation* of a compound (or an element or an ion) always involves the loss of electrons of a reacting atom. But oxidation is often easier to recognize in the form of an addition of oxygen, as in the formation of CO_2 as the product of respiration, or in the loss of hydrogen, as in the conversion of water into oxygen gas during photosynthesis. *Reduction* reactions (which always go together with oxidation processes) are the opposite: they involve a gain of electrons, or a gain of hydrogen or a loss of oxygen. During photosynthesis CO_2 is reduced as it loses oxygen.

We will encounter further types of plant reactions as we go along, and they will be explained in the respective chapters. The next section focuses on the reaction steps that plants evolved, in the process of photosynthesis.

1.6 PHOTOSYNTHESIS, KEY TO LIFE

Until about 2.5 billion years ago Earth's atmosphere contained scarcely any oxygen gas.[12,13] This started to change when some cyanobacteria (formerly called blue-green algae) evolved photosynthetic

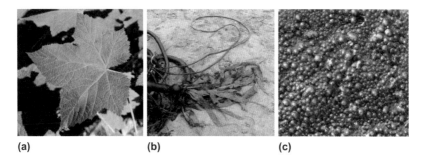

Figure 1.27 Photosynthetic green plants, algae, and cyanobacteria. (a) Leaf of thimbleberry (*Rubus parviflorus*). (b) Bull kelp (*Nereocystis* sp.), a type of algae. (c) Cyanobacteria (or blue-green algae) producing oxygen.

pigments that were capable of trapping light energy from the sun and converting it into chemical energy. This led to the synthesis of simple sugars and, as a waste product, to the formation of oxygen. Initially, the atmospheric oxygen concentration was only around 1%. Much later in Earth's history, marine algae (the phytoplankton) also evolved photosynthesis, and the oxygen content in air increased rapidly. At present our atmosphere has an oxygen content of about 21%. Green plants, algae, and cyanobacteria are the champions that are capable of performing *oxygenic* (*i.e.* oxygen-generating) *photosynthesis* (Figure 1.27).

The overall, balanced reaction equation of photosynthesis (Equation (1.5)) can be written as:

$$6CO_2 + 6H_2O \xrightarrow[\text{light}]{\text{chlorophyll}} C_6H_{12}O_6 + 6O_2 \qquad (1.5)$$

This equation is a very general and greatly simplified reaction equation as photosynthesis involves many complex reaction steps. Aside from light, water, carbon dioxide, and the pigment chlorophyll, the reactions during photosynthesis require specific enzymes, proteins and some metal ions as catalysts, too. In addition, other pigments, like the carotenoids, are also involved. They are called *accessory pigments*, and we will encounter them later in the chapter on plant pigments.

O_2 is merely a by-product of photosynthesis, albeit a most important one for life on earth. O_2 is continuously produced biologically *via* the oxidation of water. Humans and animals need the oxygen for respiration, to gain energy for further reactions. So do plants, but more oxygen is produced by plants than they use up in respiration. Some of the oxygen is transformed into ozone in the upper atmosphere to form a protective ozone layer that keeps away harmful UV radiation. The simple organic sugars that are products of photosynthesis are in turn required for the biosynthesis of all other organic plant products, like fats and oils, the plant structural materials, and the pigments, perfumes, and poisons. Humans and animals depend on basic plant products, like many sugars, fats, and amino acids that their systems cannot produce. Therefore, photosynthesis is indeed key to life on earth.

1.6.1 Light Reactions and Calvin Cycle

The reaction steps during photosynthesis can be divided into two major sets: *the light reactions* and the light-independent *Calvin cycle* (also known as *the carbon-fixation reactions*). The reactions involve several oxidation and reduction steps. A brief summary of their requirements and their products, with their interconnections, is shown in Table 1.2.

The *light reactions* (the "photo" part) occur along stacked membranes inside the chloroplast (see Figure 1.26 of a plant cell) and take place when sunlight is available. During the light reactions, photosynthetic plants, algae, and bacteria split water molecules into O_2 and H^+. Chlorophyll molecules, surrounded by proteins, capture light energy and transform it into a flow of electrons (e^-) in an oxidation reaction of water (Equation (1.6)):

$$2H_2O \longrightarrow 4e^- + 4H^+ + O_2 \qquad (1.6)$$

Table 1.2 Summary of light reactions and Calvin cycle.

Light reactions ("photo" part)		Reactions of the Calvin cycle ("synthesis" part)	
Require	*Produce*	*Require*	*Produce*
light, chlorophyll, water, and $NADP^+$, ADP produced during Calvin cycle	O_2, NADPH, ATP	CO_2 from air, and NADPH, ATP produced during the light reactions	carbohydrate, $NADP^+$, ADP

Basic Plant Chemistry Concepts

Chlorophyll is helped in this process by carotenoid accessory pigments. Electrons in the chlorophyll molecules get excited and are passed along in a series of reactions that lead to high-energy compounds, namely ATP (adenosine triphosphate). ATP is needed to power the reaction steps in the Calvin cycle. $NADP^+$ (nicotinamide adenine dinucleotide phosphate), an electron acceptor, is transformed into NADPH in a reduction reaction. NADPH is required for the reactions of the Calvin cycle as well.

For a long time it was debated whether H_2O or rather CO_2 is the source of the oxygen that is produced during photosynthesis. It was observed that some ancient hot-springs bacteria undergo non-oxygenic photosynthesis, in a reaction of CO_2 with H_2S (instead of H_2O), and form yellow sulfur (S) and simple sugars, here shown as a carbohydrate unit (CH_2O) (Equation (1.7)).[14]

$$CO_2 + H_2S \longrightarrow S + (CH_2O) \tag{1.7}$$

This observation led to the hypothesis that the source of oxygen during oxygenic photosynthesis must be from water, and not from CO_2. The hypothesis was later confirmed by incorporating the heavier ^{18}O isotopes into CO_2, and into H_2O, respectively, and following their pathways by instrumental analysis. Only if the supplied water molecules were marked with ^{18}O, did the O_2 formed during photosynthesis processes also carry the heavier isotopes.

The light-independent reactions of the *Calvin cycle* (the "synthesis" part) are named after Melvin Calvin who elucidated the reaction steps in the 1940s. They require the supply of high-energy ATP and NADPH molecules that were produced during the light reactions. The hydrogen ions and electrons, generated during the light reactions, reduce CO_2 from air to produce organic molecules in the form of simple sugars, here shown as a carbohydrate unit (CH_2O), leading to the formation of glucose, $C_6H_{12}O_6$ (Equation (1.8)):

$$CO_2 + 4e^- + 4H^+ \longrightarrow (CH_2O) + H_2O \tag{1.8}$$

These reactions also produce $NADP^+$ and ADP that are required for the light reactions, in a cycle of the reactions.

1.6.2 Special Mechanisms in Plants from Hot Climates

Plants pick up carbon dioxide through their stomata, the tiny openings at the underside of leaves and sometimes on stems, too.

But at the same time, water is lost in large quantities through the stomata. Therefore, plants that grow in hot climates need special adaptations to reduce water loss.

Most plants follow a so-called C_3 pathway. Their products of the Calvin cycle are organic three-carbon molecules that serve as starting materials for the synthesis of glucose, with six carbons. C_3 plants tend to grow well in areas where sunlight intensity is moderate, temperatures are not extreme, carbon dioxide concentrations are adequate, and ground water is plentiful.

Many plants that originate from hot climates have evolved somewhat different photosynthetic pathways that allow them to keep their stomata closed or partially closed during the hot daytime hours. This cuts down on water loss, but it also inhibits uptake of CO_2. In *C_4 plants*, the CO_2 from air is incorporated or fixed into four carbon compounds that are then transported into a different part of the cell. There the compounds break down to release CO_2 ready to be used in the Calvin cycle. Stomata can be partially closed as a supply of CO_2 is available in this fixed form, and less water is lost from the plants. C_4 plants include many grasses (Figure 1.28(a)), and also sugar cane and corn (Figure 1.28(b)), both in the grass family. *CAM plants* (for crassulacean acid metabolism) evolved a similar adaptation to hot and dry climates. Their stomata are open during the cooler night-time only, and carbon dioxide is fixed during the night by storing it as a four-carbon acid. The CO_2 is released during the day into the cell,

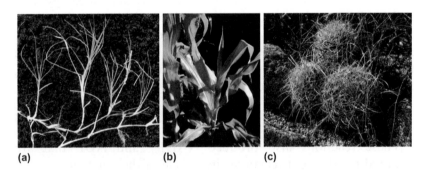

(a)   (b)   (c)

Figure 1.28 C_4 and CAM plants.
Examples of C_4 plants are (a) Bermuda grass (*Cynodon dactylon*), a drought-resistant grass with aggressive growth habits, and (b) corn (*Zea mays*), also in the grass family. (c) Barrel cactus (*Echinocactus polycephalus*), like cacti in general, is a CAM plant.

Basic Plant Chemistry Concepts

entering the Calvin cycle. The CAM pathway allows stomata to remain shut during the day, reducing loss of water. Therefore, it is especially common in plants adapted to hot and arid conditions, like in cacti (Figure 1.28(c)) or agaves.

1.6.3 Chlorophyll *a*, Main Photosynthetic Pigment

At this point let us examine the photosynthetic pigment chlorophyll more closely. The chemical structure of chlorophyll *a*, **1.1** is shown in Figure 1.29. It is a highly complex molecule. The following descriptions will explain how to read such structures.

The lines and double lines in the line structure in Figure 1.29(a) represent covalent bonds between carbon atoms. Assume a

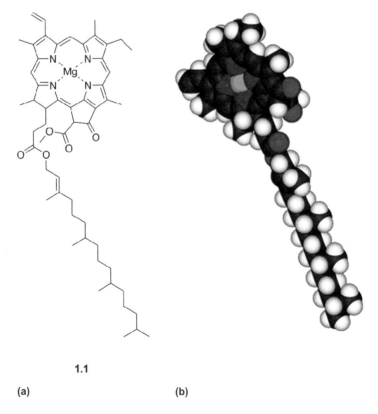

1.1

(a)   (b)

Figure 1.29 Structure of chlorophyll *a*.
(a) Line structure of chlorophyll *a*, **1.1**. (b) Space-filling molecular model of chlorophyll *a*.

carbon atom in every corner, at every intersection, and at every end of a bond. As each carbon atom needs to have four bonds, mentally attach hydrogen atoms until the four required bonds are obtained. Elements other than carbon and hydrogen, like oxygen (O, shown in red in Figure 1.29(b)) or nitrogen (N, shown in blue) are always written out, and so is the magnesium atom (shown in green) in the center of the chlorophyll molecule. A special type of bond, called a *coordinate covalent bond*, between magnesium and the nitrogen atoms is shown as broken lines. If we count all the atoms correctly and summarize them, we get the *molecular formula*: for chlorophyll *a* it is $C_{55}H_{72}O_5N_4Mg$, indeed a complex molecule.

The simple line structure (Figure 1.29(a)) shows the pattern of single bonds alternating with double bonds well. This pattern is known as *conjugated double bonds*. It is found in all colorful organic pigments. The space-filling model of the molecule in Figure 1.29(b)) displays the flat, ring-like structure, called the porphyrin ring. In its center is a magnesium atom. Bonds between metal atoms and organic molecules further intensify the absorption of light within the human visible range. Note also the long "tail" attached to the molecule, called the phytyl group: it consists of carbon and hydrogen atoms only. This makes chlorophyll water-insoluble, but soluble in non-polar media like fats and oils.

A *pigment* is generally a chemical compound that absorbs a section of the electromagnetic spectrum of sunlight (Figure 1.30) and reflects or transmits the non-absorbed portion. The segment of sunlight that is visible to humans is roughly between the wavelengths of 400 and 750 nanometers (a nanometer or nm is 0.000 000 001 meters or 10^{-9} m). When a plant pigment absorbs a portion of this visible segment and reflects or transmits the non-absorbed portion between 400 and 750 nm, then we see a colored plant part.

Chlorophyll *a*, the main photosynthetic pigment, absorbs wavelengths in the red and in the blue region of the spectrum (Figure 1.31) and reflects a combination of wavelengths that looks green to us. Other types of chlorophylls, like chlorophyll *b*, absorb different wavelengths and therefore have slightly different colors. Together with pigments like the carotenoids they act as accessory pigments. They extend the spectrum of absorbed light energy for plants. Accessory pigments like the carotenoids also serve to protect chlorophyll from an overload of electrons.

Basic Plant Chemistry Concepts

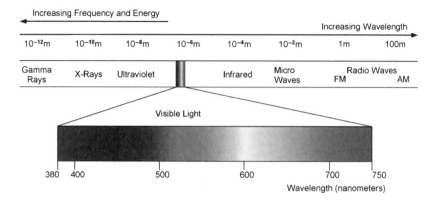

Figure 1.30 The electromagnetic spectrum.
The section of the electromagnetic spectrum that is visible to humans has wavelengths approximately between 400 and 750 nm. The shorter wavelengths of violet and ultraviolet light represent higher energy than the longer wavelengths of red and infrared. A colorful pigment absorbs a portion of the visible spectrum and reflects the non-absorbed wavelengths.

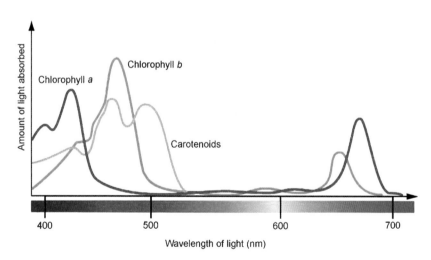

Figure 1.31 Absorption spectra of chlorophylls *a* and *b* and of carotenoids.
Chlorophylls *a* and *b* and the carotenoids absorb different portions of the spectrum, and with this different levels of energy. The combination of the non-absorbed wavelengths, reflected or transmitted to humans, appear as a green color for chlorophyll *a* and a yellow to orange color for carotenes.

Plants that do not have chlorophyll in their plant structures need special mechanisms to survive. Read about examples of such plants in Box 1.2.

> **BOX 1.2 Plants without Chlorophyll**
>
> Plants like the coral root in Figure 1.32(a) or the snow plant (*Sarcodes sanguinea*) in Figure 1.32(b) have no chlorophyll. They cannot undergo photosynthesis and are not able to produce simple sugars. Therefore, they do not have the key ingredients to synthesize other organic compounds. In order to stay alive and thrive, these plants have to obtain sugars and other nutrients from host plants or from decomposing plant materials. Once they obtain these essential nutrients, the non-green plants can compose other sugars, fats, pigments, cellulose and all other required organic compounds.
>
>
>
> **Figure 1.32** Plants without chlorophyll.
> (a) Coral root (*Corallorhiza* sp.), an orchid, and (b) snow plants (*Sarcodes sanguinea*) obtain sugars and other nutrients from decomposing plant materials.

1.7 RESPIRATION TO PROVIDE ENERGY

We are well familiar with the vital intake of oxygen during breathing which leads to the breakdown of larger molecules, like sugars and fats, in our metabolism. The process of respiration provides us with energy for all other metabolic processes, ultimately producing carbon dioxide which we exhale. The intake of oxygen is vital for plants as well. They, too, undergo respiration, and it provides them with the energy for all the other metabolic reactions. Respiration is a series of controlled reaction steps that break down larger organic molecules and create high-energy compounds, such as ATP, that then power the biosynthesis of plant molecules.

There are several biochemical pathways that plants follow during respiration, each one with multiple steps. They start out with *glycolysis*, the breakdown of sugars. In the presence of a good supply of oxygen, plants (and fungi and animals) undergo *aerobic respiration*. As carbohydrates are the main compounds that provide energy during respiration, the overall, general (unbalanced) reaction equation for aerobic respiration is shown with glucose (Equation (1.9)):

$$C_6H_{12}O_6 + O_2 \longrightarrow CO_2 + H_2O + \textbf{ATP} \qquad (1.9)$$

This equation is a very short version of a series of complex, multi-step reactions. In a complete oxidation of the reactants, the final products of aerobic respiration are carbon dioxide and water. Note that this is the reverse of photosynthesis. Respiration and photosynthesis are connected and depend on the supply of available reactants. While photosynthesis composes organic molecules of sugars, respiration is the process of producing energy by breaking them down (catabolism). Aerobic respiration produces large amounts of high-energy ATP. Some plants undergo highly increased rates of aerobic respiration at certain stages of plant development and in specific parts of the plants (Box 1.3).

BOX 1.3 Accelerated Respiration in Plants of the Arum Family

Some plants in the Arum family, like the giant corpse flower (*Amorphophallus titanum*, Figure 1.33), have evolved an alternative pathway of respiration which is activated when plants start to bloom.[15] High rates of respiration in this pathway produce large amounts of energy and can locally raise tissue temperatures by more than 10 °C compared to the ambient temperature. These plants have specialized arrangements of their blossoms, with the small flowers clustered at the bottom of a unique structure called spadix. It contains a large amount of starch and also volatile organic compounds with smells of

Figure 1.33 Accelerated respiration in plants of the Arum family. Giant corpse flower (*Amorphophallus titanum*).

rotting meat. When the corpse flower is ready to bloom, the highly accelerated respiration processes heat up the base of the spadix and cause the volatile compounds to evaporate. Their smell of carrion (highly unpleasant from a human point-of-view) attracts flies and beetles which then act as pollinators.

Anaerobic respiration is respiration in the absence of oxygen. In evolutionary terms, it is a much older form of respiration; it was the source of life's energy when oxygen was scarce. These processes produce a lot less energy in the form of ATP than aerobic respiration. Many lower organisms still live by this inefficient process, and higher organisms use it when a supply of oxygen is not available or very limited. Anaerobic conditions exist when plants are flooded or grow in mud submerged in water. Some plants like rice can undergo anaerobic respiration for a while, providing the shoots with just enough energy to grow until they reach above water, thus giving them an advantage over other plants.

Fermentation involves a partial breakdown of sugars in the absence of oxygen. Although carbon dioxide gas is still one of the products, larger organic molecules are produced as well, as shown in the following examples. The energy produced is much lower than in aerobic respiration.

Basic Plant Chemistry Concepts

In alcohol fermentation, plant carbohydrates, like glucose or starch, together with yeast and water, produce CO_2 and ethanol (C_2H_6O) (Equation (1.10)). This process is important in wine making and bread baking. (The CO_2 produced makes bread rise.)

$$C_6H_{12}O_6 \xrightarrow{\text{yeast, water}} 2C_2H_6O + 2CO_2 + ATP \qquad (1.10)$$

Lactic acid fermentation is another type of fermentation. It is used *e.g.* to make sauerkraut, by covering shredded white cabbage (*Brassica oleracea*) with salt.[16] Microorganisms, like *Lactobacillus* spp. and yeasts, induce the breakdown of sugars in cabbage. The fermentation products include lactic acid ($C_3H_6O_3$), some acetic acid ($C_2H_4O_2$), small amounts of ethanol, and CO_2.

1.8 ORGANIC MOLECULES, AN INTRODUCTION

Photosynthesis introduced organic compounds in the form of simple sugars. In everyday life, the term "organic" is used in connection with a living source, and we talk about organic gardening or organic food. In chemistry the term has a different meaning: organic chemistry is the chemistry of compounds that contain carbon.[17,18] (There are a few exceptions to this rule: for example, carbonates and CO_2 are not part of organic chemistry.) To this day, several millions of different organic compounds are known. Note that water, H_2O, is not an organic compound. But most compounds in plants are composed of organic molecules.

What leads to such a wealth of different organic compounds? Carbon atoms have the unique ability to link up with other carbon atoms, much more so than any other element. They can form chains of any length with other carbon atoms and rings of many sizes and combinations. Additional carbon atoms may be branching off from the main chains or rings. This creates infinite possibilities for potential structures. Methane gas (CH_4), produced in swamps, glucose or grape sugar ($C_6H_{12}O_6$), the product from photosynthesis, and ethanol, C_2H_6O, that we encountered as a product of fermentation, are all organic compounds. So is aspirin, $C_9H_8O_4$, the over-the-counter pain killer. A large number of organic compounds known to us are manmade.

Just a few elements assemble organic compounds: carbon (always), hydrogen (almost always), oxygen (very commonly), nitrogen (very commonly), also phosphorus and sulfur, and a couple of other non-metals. A few simple bonding rules make it possible to construct organic molecules. *Carbon* atoms always form a total of *four* bonds (with double bonds counting as two bonds, triple bonds as three). *Hydrogen* atoms always form *one* bond. *Oxygen* atoms form *two* bonds, either as two single bonds or as a double bond. *Nitrogen* atoms form *three* bonds, in the form of three single bonds, or a single and a double bond, or as a triple bond. Figure 1.34 shows some structures of organic compounds, written in different ways. As we progress through this book, organic structures will become increasingly familiar.

It is best to begin drawing organic structures by writing the carbon atoms and connecting them by their bonds. The result is the *carbon skeleton* or *carbon backbone* of a molecule. With a bit of

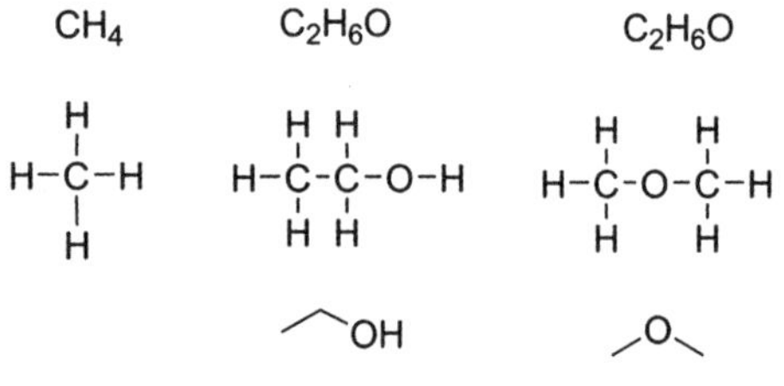

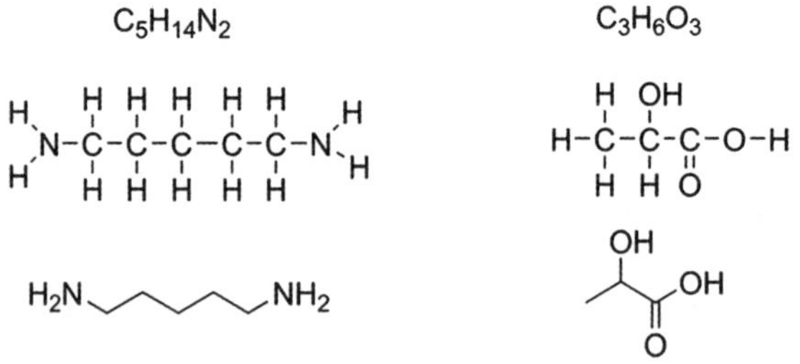

Figure 1.34 Examples of organic structures.

Each molecule is shown as a molecular formula, as a structure with all the atoms written out, and as a line structure. Methane **1.2** was introduced earlier. Ethanol **1.3** and dimethyl ether **1.4** have the same molecular formulas, but different structures. They are isomers. Cadaverine **1.5**, lactic acid **1.6**, ethanol, and dimethyl ether have different functional groups.

practice, the process can be shortened by merely drawing the bonds—leading to so-called line structures—with carbon atoms assumed in each corner, at each intersection, and at each end of a bond. This is especially helpful when drawing larger structures. Hydrogen atoms are often omitted for greater simplicity. We mentally fill in as many hydrogen atoms as are needed to give each carbon atom, or oxygen or nitrogen atom, the correct total number of bonds. Atoms like oxygen and nitrogen are always written.

The examples in Figure 1.34 illustrate these rules. Each example shows the molecular formula of a compound, the molecular structures fully written out, and the line structures. We encountered methane **1.2** earlier already. Methane, formed as a product of anaerobic decomposition of plant materials, is used as cooking gas. Ethanol or ethyl alcohol **1.3**, a product of fermentation, has the molecular formula of C_2H_6O. So does dimethyl ether **1.4**, a non-natural compound. The larger the molecules are, the more possibilities exist to assemble the atoms correctly in different structures. Compounds with the same molecular formula, but with different structures, are known as *isomers*. The formation of possible isomers contributes to the large number of different organic compounds.

Groups with atoms other than carbon or hydrogen can be bonded to the carbon skeleton. They frequently contain oxygen or nitrogen atoms. They are the *functional groups*, as they determine the chemical properties of a compound. Ethanol and dimethyl ether have such functional groups. Other functional groups are found in the compounds of cadaverine **1.5** and lactic acid **1.6**. Cadaverine is a compound with an unpleasant smell, found *e.g.* in seeds of legumes. Lactic acid was encountered as one of the compounds responsible for the sourness in sauerkraut. As we progress in our exploration of organic plant compounds, we will become familiar with typical functional groups and with some of their properties. The different functional groups even further increase the enormous diversity of organic compounds.

Organic compounds that are composed of carbon and hydrogen only are called *hydrocarbons*. They serve here as an introduction to understanding organic structures. The term "hydrocarbons" may be familiar from crude oil (petroleum), a complex mixture of hydrocarbons from ancient plant and animal materials, or from gasoline, a mixture of mostly liquid hydrocarbons. The simplest hydrocarbon is methane, shown earlier. We can write a series of hydrocarbons with increasing numbers of carbon atoms, each carbon atom forming four single bonds in a tetrahedral arrangement (Table 1.3 and Figure 1.35). Note that starting with butane (C_4H_{10}) we can write two isomers for the same molecular formula: butane and isobutane. The larger the molecules get, the more isomers are possible. Carbon can easily form double bonds to other carbon atoms, too. There are also hydrocarbons with triple bonds between carbon atoms. The bonding patterns, single or double or triple bonds, determine the shape of the molecules and with this also their chemical and biological behavior. Table 1.3 shows molecular formulas, condensed structures and line structures of representative hydrocarbons, including some examples of hydrocarbons found in plants. Molecular models, as shown in Figure 1.35, help visualize the three-dimensional shapes of molecules.

Some of the characteristic properties of hydrocarbons can be remembered by reflecting on the properties of crude oil or gasoline, both mixtures of hydrocarbons: they burn well, and they do not mix with water. Plants produce numerous different hydrocarbon compounds (Table 1.3 and Figure 1.36). Many protective layers of plants are to a large extent hydrocarbons that repel water and help plants cut down on water-loss. Hydrocarbon resins in Jeffrey pines (*Pinus jeffreyi*, Figure 1.36(a)), trees that grow in the mountains of Western North America, contain a hydrocarbon called heptane (C_7H_{16}). These resins are highly flammable mixtures. Hydrocarbons with up to about ten carbon atoms in their molecules evaporate easily as the non-polar molecules cannot form hydrogen bonds or other strong bonds between the molecules. Many plant smells are hydrocarbons that can rapidly evaporate on a warm day.

Hydrocarbons with double bonds are very common in plants. The simplest one is ethene (C_2H_4), also known as ethylene. It is a gas that acts as a ripening hormone in plants. If unripe tomatoes or avocadoes are placed in a plastic bag, the ethylene gas formed by the fruits accumulates and hastens the ripening process. Lemons (Figure 1.36(b)) contain limonene ($C_{10}H_{16}$), a hydrocarbon with a ring structure, in their peel. Limonene gives lemons their typical smell. We will encounter hydrocarbons again as colorful pigments, like β-carotene, and in the form of rubber particles in milky plant saps. A couple of hydrocarbons with triple bonds are also known from plants, *e.g.* from dahlias (*Dahlia* spp., Figure 1.36(c)), where they provide protection against insect attacks.

Table 1.3 Examples of hydrocarbons.

Name of hydrocarbon	Molecular formula	Condensed structural formula	Line structure	Comments
Methane	CH_4			In swamp gas
Ethane	C_2H_6	CH_3CH_3		
Propane	C_3H_8	$CH_3CH_2CH_3$		
Butane	C_4H_{10}	$CH_3CH_2CH_2CH_3$		
Isobutane (methyl propane)	C_4H_{10}	CH_3CHCH_3 with CH_3 branch		
Heptane	C_7H_{16}	$CH_3CH_2CH_2CH_2CH_2CH_2CH_3$		In resins of some pines
Ethene (ethylene)	C_2H_4	$H_2C=CH_2$		Ripening hormone in plants
Limonene	$C_{10}H_{16}$			Fragrance of lemons
Acetylene	C_2H_2	$HC\equiv CH$		Not found in plants
A poly-acetylene (no common name)	$C_{13}H_{10}$	$H_3C-CH=CHC\equiv C-C\equiv C$—(phenyl)		Defensive compound in dahlias

Figure 1.35 Models of some hydrocarbon molecules.
Top row (left to right): methane, butane, and isobutane. The single bonded carbon atoms lead to a tetrahedral arrangement of the bonding atoms.
Bottom row: (left) ethene, with double bond. (Right) acetylene, with triple bond.

Figure 1.36 Plants with hydrocarbons.
(a) The resins of Jeffrey pines (*Pinus jeffreyi*) contain highly flammable heptane. (b) Lemons (*Citrus* spp.) contain limonene, an oily, fragrant hydrocarbon. (c) Dahlias, like the dahlia tree (*Dahlia imperialis*), contain defensive polyacetylene hydrocarbons.

As hydrocarbons are important sources of fuels for humans, deciduous plants that grow fast and produce a high load of hydrocarbons are of great interest as potential sources for renewable energy and are keenly researched.

1.9 CONCLUSION

This chapter on basic plant chemistry reviewed key concepts in preparation for the following chapters on organic plant compounds.

Elements compose all compounds. Just a few elements, mainly carbon, hydrogen, oxygen, and nitrogen, and a few others, compose organic molecules. Mineral ions frequently serve as catalysts in plant reactions, facilitating the processes that compose organic plant compounds, all at ambient temperature and mostly in an aqueous environment. Plants, like all living beings on earth, are to a large extent aqueous systems. This means that the properties of water, like the balance (or imbalance) of hydrogen and hydroxide ions in the plant saps, greatly affect plant life. What needs to be efficiently transported in plants, has to be reasonably soluble in water. Water-insoluble compounds, on the other hand, often have protective functions. Nutrients for plants are provided by the soils they grow in, with the exception of carbon dioxide from air.

Photosynthesis delivers the basic sugars needed for the biosynthesis of all other organic plant compounds. Respiration, the breakdown of sugars, provides the energy for reactions in plants. An introduction to the structures of organic compounds included some simple rules that allow the composing of organic structures. They are a preparation for understanding and evaluating the basic families of organic compounds that are found in all living things—namely sugars, fats and amino acids—that we will encounter in the next chapter. And let us not forget that plants are capable of making these basic organic molecules from soils and air, while animals and humans need to ingest them in their plant-related diet.

BIBLIOGRAPHY AND FURTHER READING

There are many excellent texts that provide additional background on chemistry, on plant biology, and plant ecology, and that relate to many parts of this book.

Some examples are:

P. Atkins, *Atkins' Molecules*, Cambridge University Press, Cambridge, 2nd edn, 2003.

L. D. Frost, S. T. Deal and K. C. Timberlake, *General, Organic, and Biological Chemistry: An Integrated Approach*, Pearson Prentice Hall, Upper Saddle River, NJ, 2010.

M. S. Silberberg, *Chemistry, The Molecular Nature of Matter and Change*, McGraw-Hill, New York, NY, 4th edn, 2006.

N. J. Tro, *Chemistry: A Molecular Approach*, Pearson Prentice Hall, Upper Saddle River, NJ, 2nd edn, 2011.

J. McMurry, *Organic Chemistry*, Thomson Higher Education, Belmont, CA, 7th edn, 2008.

F. C. Carey and R. Giuliano, *Organic Chemistry*, MacGraw-Hill, New York, NY, 8th edn, 2010.

J. R. Hanson, *Chemistry in the Garden*, The Royal Society of Chemistry, Cambridge, 2007.

P. H. Raven, R. F. Evert and S. E. Eichhorn, *Biology of Plants*, W. H. Freeman, New York, NY, 7th edn, 2005.

J. B. Reece, L. A. Urry, M. L. Cain and S. A. Wasserman, *Campbell Biology*, Benjamin/Cummings, Menlo Park, CA, 9th edn, 2010.

J. B. Harborne, *Introduction to Ecological Biochemistry*, Academic Press, London, 4th edn, 1993.

D. L. Nelson and M. M. Cox, *Lehninger Principles of Biochemistry*, W. H. Freeman, New York, NY, 4th edn, 2005.

C. Bowsher, M. Steer and A. Tobin, *Plant Biochemistry*, Garland Science, New York, NY, 2008.

For reference on organic structures:
Merck Index, Merck & Co. Inc., Whitehouse Station, NJ, 14th edn, 2006.

REFERENCES

The end-chapter references include books on general and specific topics, as well as review articles and journal articles. While some referenced books and articles are in-depth texts, others are for popular reading.

1. J. J. R. Fraústo da Silva and R. J. P. Williams, *The Biological Chemistry of the Elements*, Oxford University Press, Oxford, 2nd edn, 2001.
2. J. B. Reece, L. A. Urry, M. L. Cain and S. A. Wasserman, *Campbell Biology*, Benjamin/Cummings, Menlo Park, CA, 9th edn, 2010.
3. P. H. Raven, R. F. Evert and S. E. Eichhorn, *Biology of Plants*, W. H. Freeman, New York, NY, 7th edn, 2005.

4. L. D. Frost, S. T. Deal and K. C. Timberlake, *General, Organic, and Biological Chemistry: An Integrated Approach*, Pearson Prentice Hall, Upper Saddle River, NJ, 2010.
5. J. B. Harborne, *Introduction to Ecological Biochemistry*, Academic Press, London, 4th edn, 1993.
6. A. R. Kruckeberg, *Introduction to California Soils and Plants: Serpentine, Vernal Pools, and Other Geobotanical Wonders*, University of California Press, Berkeley, CA, 2006.
7. P. Atkins, *Atkins' Molecules*, Cambridge University Press, Cambridge, 2nd edn, 2003.
8. B. Capon, *Botany for Gardeners*, Timber Press, OR, 3rd edn, 2010.
9. G. Sposito, *The Chemistry of Soils*, Oxford University Press, Oxford, 2nd edn, 2008.
10. G. Davies and E. A. Ghabbour, *J. Chem. Educ.*, 2001, **78**, 1609.
11. J. R. Hanson, *Chemistry in the Garden*, The Royal Society of Chemistry, Cambridge, 2007.
12. M. Leslie, *Science*, 2009, **323**, 1286.
13. P. G. Falkowski and Y. Isozaki, *Science*, 2008, **322**, 540.
14. C. Bowsher, M. Sterr and A. Tobin, *Plant Biochemistry*, Garland Science, New York, NY, 2008.
15. W. Barthlott, J. Szarzynski, P. Vlek, W. Lobin and N. Korotkova, *Plant Biol.*, 2009, **11**, 499.
16. H. McGee, *On Food and Cooking: The Science and Lore of the Kitchen*, Scribner, New York, NY, 2004.
17. J. McMurry, *Organic Chemistry*, Thomson Higher Education, Belmont, CA, 7th edn, 2008.
18. K. P. C. Vollhardt and N. E. Schore, *Organic Chemistry*, W. H. Freeman and Company, New York, NY, 6th edn, 2011.

CHAPTER 2

The Molecular Building Blocks

2.1 INTRODUCTION

This chapter provides an introduction to the families of compounds that are found in all living things: the *sugars* or *carbohydrates*, the *fats and oils*, the *amino acids*, and the *nucleotides*. They are generally known as the *primary metabolites*. Plants can synthesize them, starting with the simple sugars obtained from photosynthesis, and then connect them into larger molecules, or reassemble them into very differently structured compounds, like plant scents or plant pigments. By learning about the plant compounds we will further advance our understanding of organic chemistry. Examples of relevant plants will illustrate the different topics.

The structures of sweet *sugars* will introduce molecules that are asymmetric. Many naturally occurring compounds are composed of molecules that are asymmetric, and in order to function properly in plants (and in animals) they need to have the correct three-dimensional shapes.

While some carbohydrates feature just a few carbon atoms, others form huge molecules with a repeat pattern. These macromolecules are known as *polymers*. In everyday life, polymers may remind us of artificial materials, but polymers are a crucial component of living organisms. Plant structural materials, like cellulose and lignin, are composed of polymers, and so are proteins and nucleic acids.

The Chemistry of Plants: Perfumes, Pigments, and Poisons
Margareta Séquin
© Margareta Séquin 2012
Published by the Royal Society of Chemistry, www.rsc.org

Fats and oils are well-known for their water-repelling qualities. They are the largest and probably best-known group of compounds among the *lipids*, the water-repelling compounds in living organisms. Certain molecular characteristics let us know that a compound is water-insoluble, and also indicate whether a compound represents a liquid oil or a solid fat. Many protective, water-repelling layers in plants have large lipid components.

The most abundant macromolecules in living things are the *proteins*. These all-important polymers are composed of *amino acid* building blocks. Plants can synthesize all of the common twenty amino acids that are part of proteins. In contrast, animals and humans need to ingest some of the amino acids through their nutrition; they are called the *essential* amino acids.

Nucleic acids provide the genetic information of a plant. Ever-refined methods to study the composition of nucleic acids make it possible to compare the details of the genetic make-up of plants and help elucidate the relationships among them. This has led to quite a bit of rearranging among plant families as new relationships are recognized. The section on nucleic acids provides a brief introduction to their building blocks, the nucleotides.

Plants follow some general reaction pathways in which they make use of the primary metabolites and reassemble them into other plant molecules. A simple scheme of these pathways will be shown at the conclusion of the chapter.

2.2 CARBOHYDRATES FOR ENERGY AND STRUCTURE

Carbohydrates, also simply known as sugars, are found in all parts of plants: as plant structures in the form of cellulose, as energy storage in the form of starch in plant bulbs, or as simple sweet sugars in stems like sugar cane (Figure 2.1(a)), in roots like sugar beets, and in sweet fruits like ripe grapes (Figure 2.1(b)).[1] Flower nectars contain them, too, to attract pollinators (Box 2.1).

Carbohydrates may remind us of sweetness or of providing energy.[2] We also know how well table sugar dissolves in tea or coffee. These experiences point to common properties of carbohydrates. Sugars like glucose (grape sugar) **2.1**, fructose (fruit sugar) **2.2**, or sucrose (table sugar) **2.3** all have distinct sweet tastes, with fructose being the sweetest natural sugar (Figure 2.2). They are appreciated not only by humans, but also by some pollinators

Figure 2.1 Plants with high sugar content.
(a) Sugar cane (*Saccharum* spp.) is high in sucrose content. (Photo by Gary Tang). (b) The main sugars in ripe grapes (*Vitis* spp.) are glucose and fructose.

sipping nectar from flowers (see Box 2.1) or by animals that feed on fruits. The breakdown of carbohydrates during cellular respiration is a vital source of energy for further plant and animal processes. As for water-solubility, all sugars dissolve in water and in aqueous plant saps, with the exception of the truly large carbohydrate molecules, like starch or cellulose.

All carbohydrates are composed of C, H, and O. Their names usually end in -ose, like gluc*ose*, fruct*ose*, or cellul*ose*. As the size of the molecules allows for a large number of possible isomers, we need to write structural formulas that exactly point out which carbohydrate compound is meant. Glucose, like fructose, has the molecular formula $C_6H_{12}O_6$, and so do the sugars mannose **2.4** and galactose **2.5**, but their structures (and properties) are not alike (Figure 2.2).

Most of the time, sugar molecules form ring structures. *Simple sugars* or *monosaccharides* contain one ring ("mono" meaning "one"), whereas *disaccharides*, like sucrose **2.3**, have two rings bonded to each other (Figure 2.2). The examples are specifically shown in their β-D-form, a prefix that pinpoints exactly which ring form is meant. The bold bonds are in the foreground, pointing towards the reader.

A striking feature of sugar molecules are their many OH functional groups (known as hydroxy or alcohol groups). Water molecules can form hydrogen bonds with them and thus can surround sugar molecules and dissolve solid sugars, as shown in Figure 1.16. Many other organic plant compounds are poorly soluble in water. But if their molecules are bonded to sugar

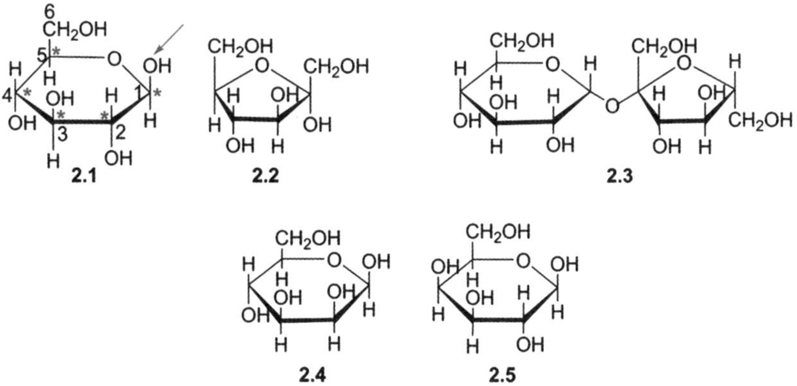

The Molecular Building Blocks

Figure 2.2 Structures of common sugars.
Glucose **2.1**, fructose **2.2**, mannose **2.4**, and galactose **2.5** are monosaccharides. Sucrose **2.3**, with two rings joined, is an example of a disaccharide. The asterisks (*) in glucose mark the asymmetric (or chiral) centers. The red arrow in glucose points to the positioning of the OH group that makes it a β-form. The bold ring bonds point to the front, towards the reader.

molecules, like glucose or other simple sugars, the additional OH groups make the entire molecules more polar and better soluble in water. They can then be transported in the plants' phloem as aqueous solutions. Non-carbohydrate molecules that are bonded to a sugar are generally known as *glycosides*. They are very common in plants, especially among plant pigments and defensive compounds. In the chapter on plant poisons for example, we will encounter glycosides in the form of the heart-active cardiac glycosides or of the cyanide-generating cyanogenic glycosides.

With sugars we encounter molecular structures that are asymmetric. Recall how carbon atoms, when they form four single bonds, do so in a tetrahedral arrangement (Figure 1.7). If a carbon atom bonds to four different atoms (or to four different continuations in a molecule), there are two different possible spatial arrangements (Figure 2.3). The two arrangements are not alike and cannot be superimposed. They are like mirror images, or *enantiomers*, to each other. Compare with a left and a right hand. The two hands are mirror images of each other, but distinctly different, and a left-handed glove will not fit a right hand. Hands are asymmetric. Figure 2.3 shows the two mirror arrangements of an asymmetric carbon (a) as wedge-and-dash structures and (b) as molecular

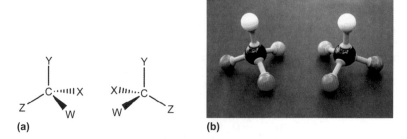

Figure 2.3 Mirror images of asymmetric carbon centers.
Mirror images, or enantiomers, of a carbon atom bonding to four different atoms (or to four different continuations of the molecule) are represented (a) as wedge-and-dash structures and (b) as models. The two enantiomers are not superimposable.

models. The carbon atom at the center of the tetrahedron is called an asymmetric or *chiral center* (from Greek for "handed"). Many compounds in plant and animal systems have asymmetries in their molecules, sometimes with several chiral centers. Their three-dimensional shapes determine if and how the molecules interact with other asymmetric molecules. Asymmetric compounds and their mirror images form types of isomers that are generally known as *stereoisomers*; they are isomers that are only different in their spatial arrangements. Very commonly, only one specific stereoisomer can function in a biochemical reaction. Examples of stereoisomers among simple sugars are described in the next paragraph.

The ring structure of glucose **2.1**, as shown in Figure 2.2, has five chiral centers. Each one is marked with an asterisk (*). The six-membered ring structure of glucose actually exists in two versions. One has the OH group at carbon number 1 above the ring, as shown (the β-form); the other one has this OH group below the ring (the α-form). (The ring can open and close at carbon number one, because of its special functional groups, and can reclose in either the β-form or the α-form.) While both forms are still glucose and the distinction may seem unimportant, the position of this OH group is of great importance when glucose links to other glucose units in larger carbohydrate molecules, as will be shown later in the examples of starch and cellulose. Note that the ring structures of mannose **2.4** and galactose **2.5** (Figure 2.2) differ in the positioning of one or two OH groups bonded to other chiral carbons. They are stereoisomers of glucose. Mannose

The Molecular Building Blocks

and galactose are different sugars, with different reactivities in plants—and with different sweetness to us. Apparently, our sensory organs are asymmetric as well and interact differently with the unlike stereoisomers. Box 2.1 further illustrates the role of different sugars as components of flower nectars.

BOX 2.1 Flower Nectars and Their Composition

Flowers have evolved numerous, diverse ways to attract pollinators. Many blossoms offer a sweet, sugary solution, called nectar, and attract bees, butterflies, hummingbirds, or small bats, occasionally even small mammals, as illustrated with the plant examples in Figures 2.4(a) and 2.4(b). Glucose, fructose, and sucrose are found in all nectars.[3,4] However, their relative amounts in floral nectars vary quite a bit. The specific sugar compositions of nectars seem to be especially attractive to certain pollinators. Nectars also contain low concentrations of proteins that provide an important source of nutrition for animals like butterflies that mainly feed on nectar.

Some flower nectars contain toxic compounds that may be selective to pollinators accustomed to them. These toxins can be as simple as a different type of sugar, like galactose (in some tulips) or mannose (Figure 2.2). Honeybees do not have the necessary

(a)   (b)   (c)

Figure 2.4 Blossoms with nectar.
(a) Small native bees try to get directly to the sugary nectar by boring holes into the blossoms of manzanita (*Arctostaphylos* sp.). (b) The low-growing, nectar-rich blossoms of *Protea cordata*, from South African Cape Province, provide easy access to small rodents that act as pollinators. (c) Nectars from *Rhododendron* spp. contain toxic grayanotoxins.

> enzymes to digest them and fall ill. The toxins can also be compounds with non-carbohydrate structures, like the so-called grayanotoxins in the nectars of rhododendrons and azaleas (Figure 2.4(c)).[5] Honey from these plants contains the toxins, too, and needs to be avoided.

Glucose is the most common carbohydrate in plants. It has numerous roles in plant life. Earlier we encountered glucose as the product of photosynthesis, as major source of energy during respiration, and as a key compound for the biosynthesis of all other organic plant compounds. Glucose contributes to flower nectars and enhances water-solubility of plant compounds as part of glycoside molecules. In addition, glucose or sucrose can serve as freeze protection in plants. An elevated sugar content in plant saps can help plants tolerate low temperatures. Another major role of glucose is as the building block (or *monomer*) of huge molecules with a repeat pattern, called *polymers*. Thousands of glucose monomers link up in specific bonding patterns to form the polymers of starch and cellulose. The ways in which the glucose molecules are bonded to each other determine the shapes of the polymers and how enzymes interact with them.

Starch is the main form of energy storage in plants. It consists of a combination of two related polymers: amylose and amylopectin. In both of them, the glucose units are connected by so-called α-glycosidic linkages **2.6**. Note that carbon atoms C1 and C4 are chiral carbons (Figure 2.5). The naming of "α-glycosidic" points out the specific bonding around the chiral centers that are part of the links.

Starch is much more stable than glucose; it can withstand heat or cold far better and does not get oxidized easily. While glucose molecules rapidly dissolve in water, affecting the osmotic pressure in the cell, starch is not soluble in cold water, due to the large size of the molecules. Therefore starch is an ideal form of storage of sugars in plants. When needed, plants can break down starch to glucose, with the help of special enzymes. Underground plant structures, like bulbs, tubers, or corms (Figure 2.6), contain plenty of starch to store energy. Green leaves usually have starch granules as a quick source of energy. Seeds all contain starch as energy supply for the future seedlings. Note that many foods for humans, like rice, corn, wheat, or potatoes (Figure 2.6(b)), are plant seeds or tubers with a

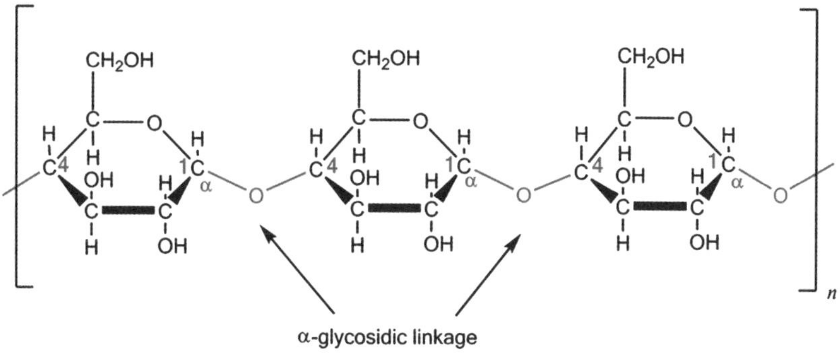

The Molecular Building Blocks 55

![Structure showing three glucose units linked by α-glycosidic linkages]

α-glycosidic linkage

2.6

Figure 2.5 Bonding of glucose units in starch.
Thousands of glucose monomers link up in specific fashion, in α-glycosidic linkages, to form starch polymers. The bold bonds point to the front, towards the reader.

(a)   (b)   (c)

Figure 2.6 Bulbs, tubers, and corms for storage of starch.
(a) Tulip and daffodil bulbs. (b) Potato plant (*Solanum tuberosum*) with tubers, *i.e.* potatoes. (c) Corms of Montbretia plants (*Crocosmia* cultivar).

high content of starch. Humans have the appropriate, correctly shaped enzymes in their digestive systems that can break down starch to glucose.

Plants with underground starch reservoirs, in the form of bulbs, tubers, or corms, are particularly common in parts of the world with Mediterranean climates. These climate zones, found in the Mediterranean region, as well as in South Africa, California, Western Australia, and in parts of Chile, are characterized by maximum rainfall in winter and long months of drought during summer. Plants that can survive a lengthy dry season need special adaptations. Many plants in Mediterranean climates die back during the dry spells and conserve starch (and water) in underground plant organs for the next growing season.

2.3 STRUCTURES THAT KEEP PLANTS UPRIGHT

Plants need to reach for light, and many need to stand out for pollinators. What keeps sunflowers, towering redwood trees (*Sequoia sempervirens*), and green scouring rush plants (*Equisetum arvense*) standing upright (Figure 2.7)?

Cellulose, another polymer of glucose, is the structural material in green plant stems and leaves (Figure 2.8). When looking at a segment of a cellulose molecule **2.7**, note how the glucose units in cellulose molecules are joined differently compared to starch: in cellulose, the glucose monomers are linked by β-glycosidic linkages. Humans do not have the correct enzymes to break down these polymers into glucose units. Cellulose is merely fiber for us. Animals like cows or deer (the ruminants), on the other hand, have special bacteria in their digestive systems, with the correct enzymes to break down cellulose into glucose. These animals can then make use of glucose as a source of energy.

While the huge size of the polymeric cellulose molecules does not allow them to dissolve in water, cellulose soaks up water easily, due

Figure 2.7 Reaching for light.
(a) Wild sunflowers (*Wyethia mollis*). (b) Coast redwood trees (*Sequoia sempervirens*). (c) Horse tails or scouring rush (*Equisetum arvense*).

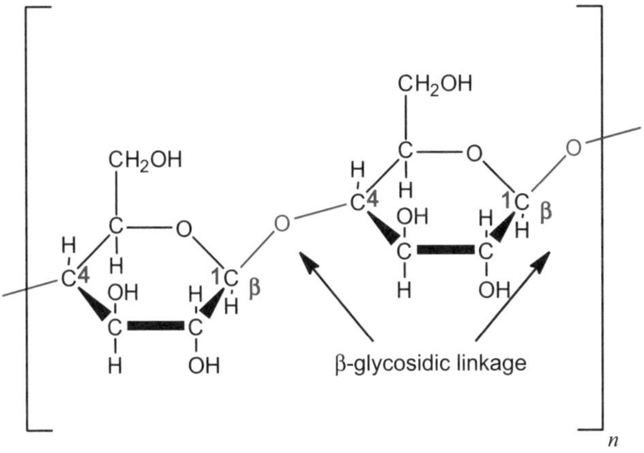

The Molecular Building Blocks 57

2.7

Figure 2.8 Bonding of glucose units in cellulose.
Glucose monomers in cellulose, unlike in starch, are connected by β-glycosidic linkages.

to its many OH groups. (Just think how a cotton shirt, cotton being mostly cellulose, stays wet for a long time after a rain.)

When plant parts turn woody as they age, they form lignins in their bark.[6,7] Leaf veins, roots, and fruits can contain them, too. Lignins provide support, but also resistance to pathogenic organisms. Lignins are composed of complex polymers that contain a great variety of monomers. In woody structures of trees, the variation in composition of lignins lead to different packing of their polymeric structures. This in turn determines the properties of a type of wood, like hardness or softness.

Figure 2.9 shows two common monomers in lignins: coniferyl alcohol **2.8** and sinapyl alcohol **2.9**. Both have benzene rings (shown in red), *i.e.* six-membered rings that are fully conjugated. Compounds that contain this type of ring are called *aromatic* (even if they have no smell at all.). Aromatic structures are very stable and have their own chemical reactions. We will encounter many aromatic rings in the plant molecules shown in this book. Many monomers in lignins have a C_6-C_3 (or phenyl propanoid) structure, so-called because the aromatic rings, with six carbons in them, each have three-carbon side chains attached. Phenyl propanoid units are quite common in plant compounds. Try to find these units in the lignin segment shown in **2.10**.

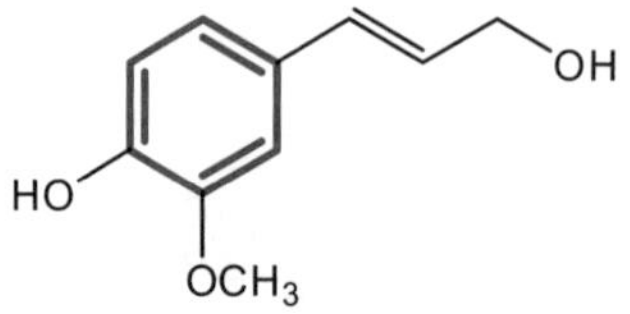

Figure 2.9 Structures of lignin.
Coniferyl alcohol **2.8** and sinapyl alcohol **2.9** are common structural units in lignin. Both are aromatic (with the benzene rings shown in red color) and contain C_6-C_3 (phenyl propanoid) units. The section of lignin **2.10** shows some of these units.

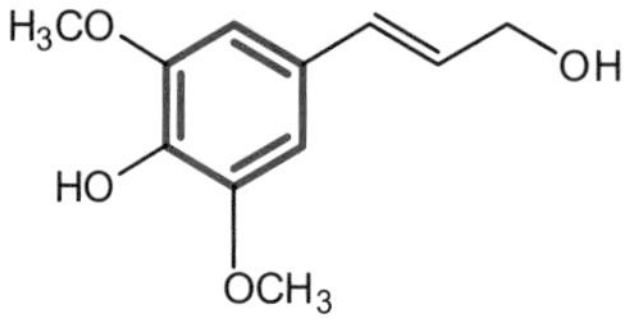

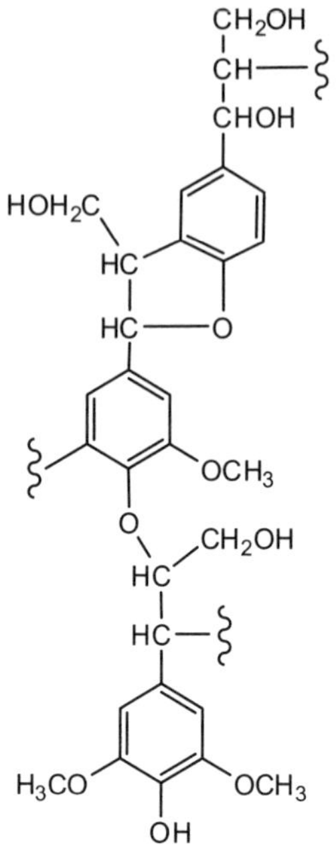

Lignins form protective layers in plants. While many fungi decompose cellulose, by breaking down the glycosidic bonds, few organisms are capable of breaking down lignin. When certain leaves decompose on the ground, they often show a lacework of the lignified veins, while the less stable cellulose portion of the leaf has decayed, as shown in Figure 2.10.

Horsetails or scouring rushes (*Equisetum* spp.) are ancient plants that evolved mechanisms to convert silicon-containing compounds from the soil into quartz or silica (SiO_2).[8] The plants feel rough when their stems are rubbed, because of the silica deposited as a coarse layer on the outside of the stems. The rough layers help the plants to stay upright and also provide them with increased resistance against animals that intend to feed on them. The name

The Molecular Building Blocks

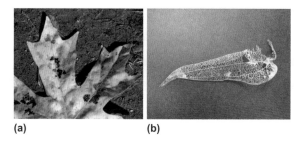

(a)   (b)

Figure 2.10   Cellulose and lignin.
(a) A leaf of bigleaf maple (*Acer macrophyllum*) in fall, showing beginning decay of the cellulose, with the more resistant lignified veins left. (b) The vein skeleton of a decomposed leaf of *Lapageria rosea*.

"scouring rush" alludes to the use of these plants for scouring pots and pans in former times.

2.4 FATS AND OILS FOR ENERGY AND PROTECTION

From our daily lives we know that oils and water do not mix. Compounds from living organisms that do not dissolve in water are generally known as *lipids*. They are non-polar, hydrophobic substances. Typically, lipid molecules have large hydrocarbon sections, with few (if any) polar groups like OH.

Fats and oils are the largest family of lipids, and we will focus on them here. Many plant seeds are rich in fats and oils.[9] The breakdown of their fats and oils during respiration serves as a source of high energy and allows the future seedlings to grow. Peanuts (Figure 2.11(a)), sunflower seeds (Figure 2.11(b)), and corn are all seeds that are high in oil content, with olives being some of the few fruits that supply oils (Figure 2.11(c)). They are all used to supply energy in our own nutrition. Plants are the source of essential fats and oils that humans and animals need in their diet.

Most fats and oils are *triglycerides* (Figure 2.12). We can understand their structures as formally assembled from glycerol and three fatty acid groups. Glycerol **2.11** is an alcohol with three OH groups. Fatty acids are long hydrocarbon chains with a carboxylic acid (COOH) functional group attached. Examples of fatty acids are palmitic acid **2.12**, oleic acid **2.13**, and linoleic acid **2.14**, all shown as line structures as well as space-filling models in Figure 2.13. Carboxylic acid groups can react with the alcohol

Figure 2.11 Plant sources for plant oils.
(a) Raw peanuts (*Arachis hypogea*). (b) Sunflower (*Helianthus annuus*) with seeds. (c) Olive tree branch (*Olea europaea*) with olives.

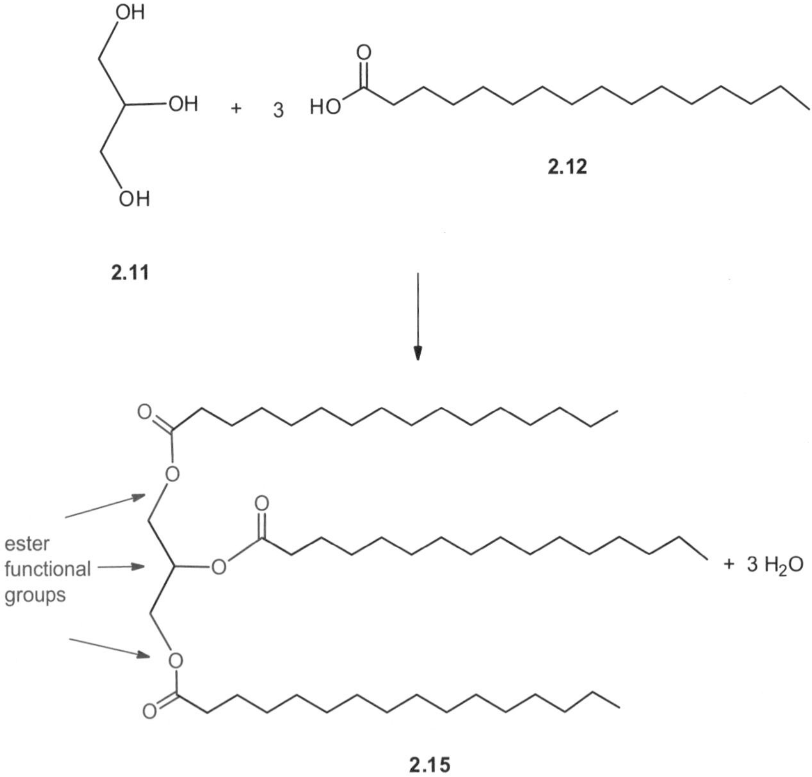

2.11

2.12

ester functional groups

2.15

+ 3 H$_2$O

Figure 2.12 Glycerol and fatty acid forming a triglyceride.
Glycerol **2.11** and three palmitic acid molecules **2.12** are shown forming a triglyceride or fat **2.15**, with three ester functional groups.

The Molecular Building Blocks 61

groups of glycerol (and do so during metabolism) by losing water and forming a new functional group, the *ester* group; it is pointed out in Figure 2.12. Three fatty acids, either the same or different ones, can react with glycerol, and form a triglyceride **2.15** which is a fat or an oil. Note the three ester groups. The long hydrocarbon chains in natural fats or oils always have an even number of carbon atoms (like 14, 16, 18, or 20), because their biosynthesis in plants involves an assembly from two-carbon units, as will be shown later in this chapter.

The molecular structures of triglycerides provide quite a bit of information about the nature of a fat or oil. Not only do the long hydrocarbon parts in the molecules show that these compounds are hydrophobic. Their molecular structures inform us also whether they represent a liquid oil or a solid fat at ambient temperature. Palmitic acid **2.12** has single bonds only in the hydrocarbon chain that are able to rotate freely (Figure 2.13). When these long hydrocarbon strands become part of a triglyceride molecule, as in **2.15**, they can pack densely due to their regular shapes. This results in a compound that is likely to be solid at room temperature. Solid fats are mostly *saturated*, meaning that they have few, if any, double bonds. Animal fats are commonly saturated fats.

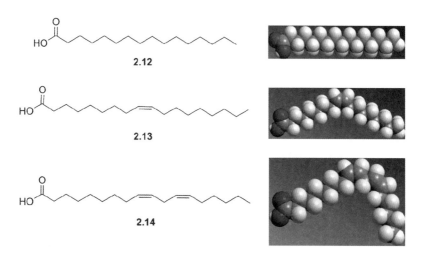

Figure 2.13 Structures of saturated and unsaturated fatty acids.
Palmitic acid **2.12** is a saturated fatty acid. The introduction of double bonds leads to bent hydrocarbon chains, as shown in the space-filling models of oleic acid **2.13** and linoleic acid **2.14**.

Figure 2.13 compares the structure of palmitic acid with those of fatty acids that have double bonds in their side chains, *i.e.* that are *unsaturated*. Note that the introduction of a double bond into the hydrocarbon strands produces a bend in the chain, as in oleic acid **2.13** with one double bond, or in linoleic acid **2.14** with two. Bent hydrocarbon chains cannot arrange themselves as compactly as saturated chains in triglyceride molecules. As a consequence, triglycerides with unsaturated hydrocarbon chains have lower melting points and are likely to be liquid oils at room temperature. Plant oils in general are unsaturated and tend to be liquid. Many plants that can survive cold climates have a high degree of unsaturation (*i.e.* many double bonds) in the hydrocarbon chains that are part of their cell membranes. This allows membranes to maintain a semiliquid, functioning state, even at low temperatures.

Unsaturated hydrocarbons are chemically much more reactive than saturated ones. Moisture and oxygen from air can react with the double bonds in oils, forming new, mostly unwanted, compounds. For that reason, unsaturated plant oils, like olive oil, turn bad or rancid faster than the more saturated animal fats like lard.

Another important aspect of triglyceride molecules relates to the positioning of atoms around the double bonds in their hydrocarbon chains. Carbon-carbon double bonds are rigid. Whenever a molecule contains such a structure, the two carbons forming the bond and the atoms next to them form a flat (or planar) arrangement, with no possibility of rearranging the atoms without breaking a bond (Figure 2.14). If different groups are attached to the carbon-carbon double bonds (as indicated by the colors green

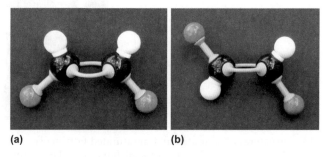

Figure 2.14 *Cis* and *trans* arrangements around carbon-carbon double bonds. (a) *Cis* arrangement. (b) *Trans* arrangement.

The Molecular Building Blocks

and white in the models in Figure 2.14), the attachments can either be on the same side (*cis*) of the double bond, as shown in Figure 2.14(a), or at opposite sides (*trans*), as shown in Figure 2.14(b). They are *cis/trans* isomers and represent another type of stereoisomer. (Earlier we encountered stereoisomers of compounds with chiral carbons). The molecules of plant oils always have a *cis* arrangement around their double bonds. Recheck the structures of oleic and linoleic acid in Figure 2.13 to verify their *cis* arrangements.

The expressions of "*trans* fats" and "saturated fats" may sound familiar from food items. Labels on nutritional products that contain fats and oils usually list the contents of unnatural "*trans* fats." They mention as well the percentage of "saturated" and of "polyunsaturated" fats, the prefix "poly" relating to many double bonds.

Protective waxes on plant leaves, the plant hydrocarbons described in the previous chapter, volatile plant odors, and plant steroids that we'll encounter later are all lipids. Box 2.2 specifically describes some plant layers that are highly water-repellent and thus protect plants.

BOX 2.2 Protective Water-Repellents

Many lipids have protective functions in plants. Hydrocarbons in the form of resins protect the bark of pine trees, especially when trees are wounded. Resinous layers also cover leaves of plants and help them cut down on water loss during long dry spells. Figure 2.15(a) shows a shiny, resinous leaf of Yerba Santa (*Eriodictyon californicum*), a shrubby plant of Western United States. The later chapter on plant defenses will elaborate more on these protective resins.

Waxes on fruits form protective coatings on many fruits. You may have made an apple shiny by rubbing its surface. Apples, plums, and oranges fresh from the tree have thin, whitish wax layers. The waxes reduce drying of the fruits and resist pest attack. Waxes are mostly esters of fatty acids and alcohols with long hydrocarbon chains that together give them non-polar properties.

Figure 2.15 Plants with protective water-repelling layers.
(a) Resinous leaf of Yerba Santa (*Eriodictyon californicum*).
(b) Nopal cactus (*Opuntia* sp.) with thick cuticles. (c) Bark of cork oak (*Quercus suber*).

Strongly hydrophobic layers, called *cuticles*, cover leaves and young shoots of many plants.[10] Desert plants especially, like cacti, have thick cuticles that help them survive the dry climate (Figure 2.15(b)). The waxy layers repel water from the plant surfaces and keep them from drying out. Thick cuticles also provide structural support and a barrier against pathogenic organisms. The hydrophobic, water-proofing layers on leaves keep water away from delicate cellulose that could soak up water and then start to rot. Cuticles consist mostly of *waxes* and a complex polymer, called *cutin*. The latter is a polymer with large lipid sections. Many ester groups connect the hydrocarbon chains and form also crosslinks between the chains. Cutin is highly unreactive and provides excellent protection against physical and chemical injuries from the environment.

The trunks of cork oaks (*Quercus suber*) are covered by thick layers of cork (Figure 2.15(c)). This highly water-repellent material, used to make corks on wine bottles, is composed of another complex, polymeric substance, called *suberin*.[11] It is formed in secondary tissues in plants. It is present in the underground parts of vegetables, as in the peel of potatoes, and also in the bark of trees. Sections of the polymeric suberin molecules resemble the ester structures of waxes; others remind us of the aromatic structures of lignins. Figuring out the structures of suberin and cutin is challenged by the insolubility of their highly inert polymeric structures, and research on them is still in progress.

The Molecular Building Blocks 65

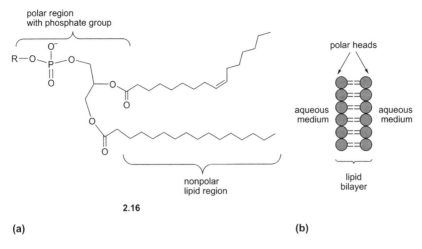

Figure 2.16 Phospholipids.
(a) A general structure of a phospholipid, with the polar phosphate region and the nonpolar lipid region. (b) A simple drawing of a segment of a lipid bilayer, with the polar heads directed towards the aqueous medium.

Many large molecules in living systems have both non-polar and polar parts: the non-polar lipid sections can interact with other lipids, while the polar parts of their molecules can hydrogen-bond with water. Phospholipids are important examples of this type of lipids. They are essential components of biomembranes that surround and compartmentalize structures in living cells.[12] These membranes are selectively permeable, *i.e.* allow passage only of selected ions and molecules. Their polar regions are oriented towards the polar aqueous medium and the lipid tails towards the inside of the membrane. A general example of a phospholipid **2.16** is shown in Figure 2.16(a)), with the fatty acid groups constituting the non-polar lipid region and the phosphate group the polar region. "R" represents a variable group. Phospholipids compose the lipid bilayers of membranes, as shown in the simple drawing in Figure 2.16(b).

2.5 PROTEINS WITH MANY FUNCTIONS

2.5.1 Amino Acids, the Building Blocks of Proteins

Hundreds to thousands of amino acid units link up to compose the polymers called *proteins*. These huge molecules are part of all plant (and animal) cells and of all cell components. They are the most abundant polymers in living organisms and have highly diverse structures, shapes, and functions. Some act as structural proteins and provide support in cell membranes. Other proteins function as enzymes and act as highly specific catalysts in biochemical reactions. Yet others provide storage; when broken down they are a quick source of amino acids in times of need. Plant seeds that must provide nutrients for the future seedlings are rich in proteins. Beans and nuts are examples (Figure 2.17(a) and (b)). Again, it is no coincidence that we use many plant seeds as staple foods, like rice, wheat, beans, and corn, to obtain essential amino acids and proteins for our own nutrition. Some proteins may also act as plant defenses in the form of toxic plant compounds. An example is ricin toxin, a highly poisonous, water-soluble protein from the castor bean plant (*Ricinus communis*) (Figure 2.17(c)).

Peptides are also composed of amino acid units but are smaller molecules than proteins, with anywhere from two to about a hundred amino acid units linked up. We first focus on the specific structures of *amino acids*, the building blocks of peptides and proteins.

A set of only twenty different amino acids, the same for all living things, composes the thousands of different proteins and peptides. Amino acids contain nitrogen in addition to carbon and oxygen in their molecules. Plants can synthesize all twenty amino acids, starting from glucose and a nitrogen source, like ammonia or nitrate from the soil. (Animals and humans on the other hand have

Figure 2.17 Plant sources of proteins.
(a) Various types of beans. (Display at University of California Botanical Garden, Berkeley.) (b) Mixed nuts. (c) Castor bean plant (*Ricinus communis*). (Photo by Joan Hamilton.)

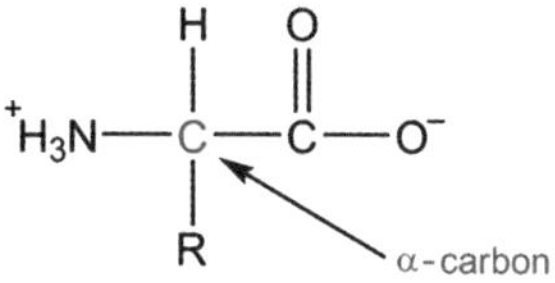

The Molecular Building Blocks

$$^+H_3N-\underset{R}{\overset{\overset{H}{|}}{C}}-\overset{\overset{O}{\|}}{C}-O^-$$
α-carbon

Figure 2.18 General structure of an α-amino acid.

to obtain eight amino acids through their nutrition. They are called *essential* amino acids.) As the name implies, amino acids have an acid part, which is a carboxylic acid (COOH) group, and also an amino (NH_2) functional group. The amino group is basic and can accept hydrogen ions, namely from the acidic carboxylic acid group which, as an acid, can donate hydrogen ions. Figure 2.18 shows a general structure for amino acids, with "R" representing the varying side chains in different amino acids and with the two functional groups in their ionic state (note the charges). The pH of plant saps affects how ionized the functional groups are. Most of the amino acids in plants (and all of those that are important to humans) have the two functional groups attached to the same carbon atom, called the α-carbon, and therefore are known as *α-amino acids*.

Figure 2.19 shows examples of amino acids that are part of proteins: the amino acids glycine **2.17**, alanine **2.18**, phenyl alanine **2.19**, methionine **2.20**, aspartic acid **2.21** and lysine **2.22**. For simplicity, they are shown in their nonionized forms. The side chains, described as "R" earlier, can contain sulfur, as in methionine **2.20**; others have additional acid or amino groups, as in aspartic acid **2.21** or in lysine **2.22**. Note that α-amino acids, with the exception of glycine, have four different groups bonded to their α-carbon atom. This means that they have an asymmetric or chiral center at the α-carbon, marked with an asterisk (*) in the examples in Figure 2.19. Interestingly, only one of the mirror images, the left-handed or L-form, is part of proteins found in life on Earth. Figure 2.20 shows an illustration of the enantiomers of the amino acid alanine, with the left-handed (natural) L-form and the right-handed D-form. Chapter 6 on "Plants and People" will elaborate on essential amino acids in human nutrition.

Amino acids provide the nitrogen in many biological plant reactions that compose other nitrogen-containing plant compounds,

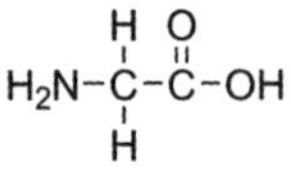

$$\underset{2.17}{\overset{H}{\underset{H}{H_2N-\overset{|}{\underset{|}{C}}-\overset{O}{\overset{||}{C}}-OH}}} \quad \underset{2.18}{\overset{H}{\underset{CH_3}{H_2N-\overset{|}{\underset{|}{C^*}}-\overset{O}{\overset{||}{C}}-OH}}} \quad \underset{2.19}{\overset{H}{\underset{CH_2-C_6H_5}{H_2N-\overset{|}{\underset{|}{C^*}}-\overset{O}{\overset{||}{C}}-OH}}}$$

2.17 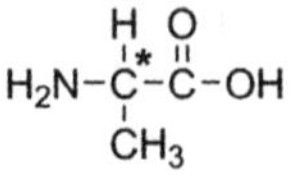   2.18 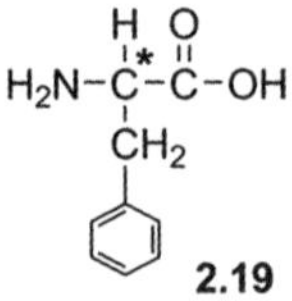   2.19

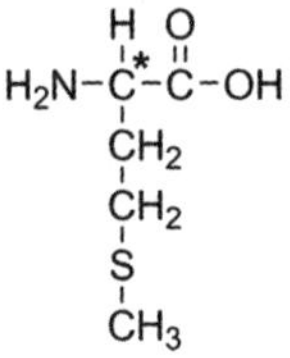

Structure 2.20: H$_2$N-C*H(CH$_2$CH$_2$SCH$_3$)-COOH (methionine)

Structure 2.21: H$_2$N-C*H(CH$_2$COOH)-COOH (aspartic acid)

Structure 2.22: H$_2$N-C*H(CH$_2$CH$_2$CH$_2$CH$_2$NH$_2$)-COOH (lysine)

2.20 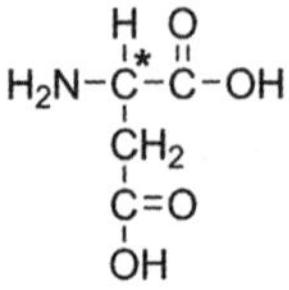   2.21 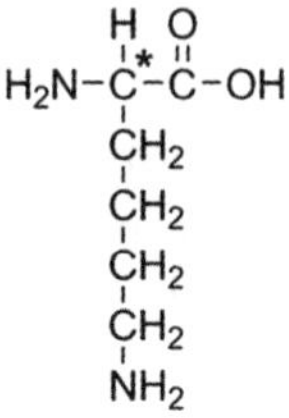   2.22

Figure 2.19   Examples of α-amino acids that compose proteins. Glycine **2.17**, alanine **2.18**, and phenylalanine **2.19**, methionine **2.20**, aspartic acid **2.21**, and lysine **2.22** are common amino acids that are part of peptides and proteins. With the exception of glycine, amino acids have a chiral center (marked with *) at the α-carbon.

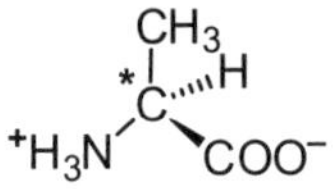

L-Alanine 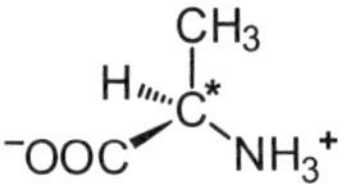   D-Alanine

Figure 2.20   Enantiomers of the amino acid alanine.

like the bases in nucleotides of nucleic acids shown later in this chapter, or the alkaloids, the family of organic plant bases that will be introduced in the chapter on plant defenses. The biosynthesis of amino acids in living organisms and their assembly into proteins have direct links with the genetic information stored in the nucleic acids.

2.5.2 Peptides and Proteins

Now that we have learned about the structures of amino acids, we can study how they condense to form the larger molecules of peptides and the polymeric proteins. Amino acids can link up by losing water and forming peptide bonds, -CONH- (shown in red),

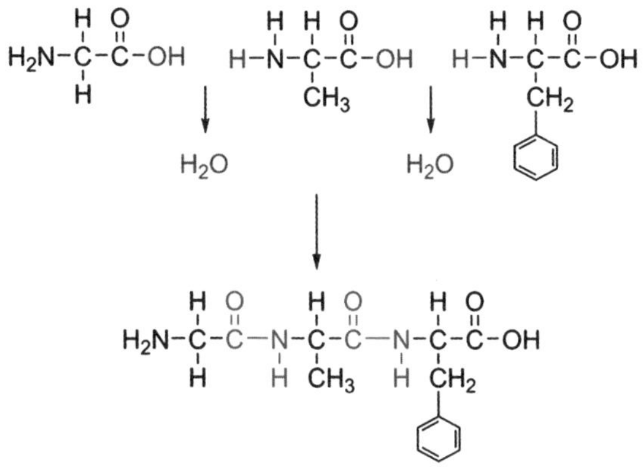

The Molecular Building Blocks

$$H_2N-\underset{H}{\underset{|}{\overset{H}{\overset{|}{C}}}}-\overset{O}{\overset{\|}{C}}-OH \quad H-\underset{|}{\overset{H}{N}}-\underset{CH_3}{\underset{|}{\overset{H}{\overset{|}{C}}}}-\overset{O}{\overset{\|}{C}}-OH \quad H-\underset{|}{\overset{H}{N}}-\underset{CH_2-C_6H_5}{\underset{|}{\overset{H}{\overset{|}{C}}}}-\overset{O}{\overset{\|}{C}}-OH$$

$$\downarrow H_2O \qquad \qquad \downarrow H_2O$$

$$H_2N-\underset{H}{\underset{|}{\overset{H}{\overset{|}{C}}}}-\overset{O}{\overset{\|}{C}}-\underset{H}{\underset{|}{N}}-\underset{CH_3}{\underset{|}{\overset{H}{\overset{|}{C}}}}-\overset{O}{\overset{\|}{C}}-\underset{H}{\underset{|}{N}}-\underset{CH_2-C_6H_5}{\underset{|}{\overset{H}{\overset{|}{C}}}}-\overset{O}{\overset{\|}{C}}-OH$$

Figure 2.21 Formation of peptide bonds.
Amino acids condense by losing water (shown in blue) and forming peptide bonds (highlighted in red).

between the amino acid units (Figure 2.21). Any number and combination of amino acid units can be linked this way into any length of peptide or protein required.

There are several levels to look at peptide and protein structures (Figure 2.22).[13] The order or sequence in which amino acids are linked is called the *primary structure* of a peptide or protein (Figure 2.22(a)). Can you recognize the primary structure of the peptide in Figure 2.21? The genetic information from the DNA in the nucleus of the cells ultimately determines the primary structures of peptides and proteins.

As we are dealing with very large molecules, hydrogen bonding can occur within the molecules. It determines the shape of proteins or peptides, leading to helical forms and more sheet-like structures, called the *secondary structure* (Figure 2.22(b)). On a higher level of order, longer-range spatial arrangements within these huge molecules further influence the shapes of proteins and can lead to more roundish, globular structures, or to fibrous structures. This is called their *tertiary structure* (Figure 2.22(c)). There are even *quaternary structures*, composed of two or more separate protein subunits (Figure 2.22(d)). The proper functioning of a protein is determined by its three-dimensional arrangement. It is very often an asymmetric shape that interacts specifically with other asymmetric molecules, most importantly in enzyme interactions.

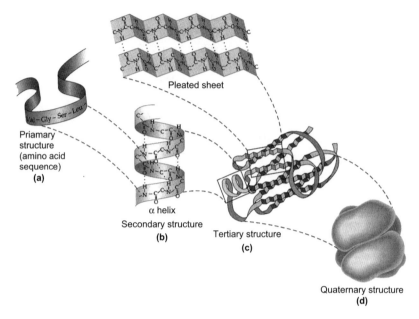

Figure 2.22 The four levels of protein structure.
(a) Primary structure (amino acid sequence). (b) Secondary structure due to hydrogen-bonding, forming α-helix or β-pleated sheet. (c) Tertiary structure. (d) Quaternary structure. (Image from Neil A. Campbell, Biology, 4th edn, © 1996, p. 81. Reprinted by permission of Pearson Education, Inc., Upper Saddle River, NJ.)

Peptides and proteins are fragile structures. Heat, high content of metal ions in the soils or pH extremes will destroy their secondary, tertiary, and quaternary structures as hydrogen bonds within the protein molecules are disrupted. This leads to the loss of the correct folding patterns of the molecules—and to a loss of proper function. The destruction of the native form of a peptide or protein is known as *denaturation*.

2.6 NUCLEIC ACIDS AND GENETIC INFORMATION

How do plants inherit the shape of leaves or the color of their flowers from their parent plants? How is the information about plant smells or about defensive compounds given to the offspring? Nucleic acids in the nuclei of plant cells store the genetic information. Research on nucleic acids and genetics is a field that has

The Molecular Building Blocks

expanded explosively over the last decades. Many plants are newly assigned to other plant families as information about their genetic make-up is obtained, and as investigations provide information about the relationships between plants and their evolution. Genetic research is also of intense interest in the search for higher-yielding crop plants, in ever-more challenging environmental conditions, to nourish ever-increasing human populations. The later chapter on "Plants and People" will go more into details about genetically modified plants.

The previous section addressed the abundance of proteins and their numerous and highly vital functions, showing how the structures of proteins determine their functions in plants. These structures in turn are determined by the sequence of amino acids in the proteins. The information about the primary structure of proteins is programmed by sections in deoxyribonucleic acid or DNA.

Deoxyribonucleic acids (*DNA*) are polymers in the cells' nuclei that carry the genetic material from the parent organisms.[14] But the information about all the cells' activities, like the types of proteins, is not directly transmitted. Instead, the coded instructions from the DNA are read by a second type of nucleic acid, the ribonucleic acid or *RNA*. So-called messenger RNA then transports the instructions out of the nucleus and interacts with the protein-synthesizing machinery in the cell that translates the genetic code into proteins. The actual protein synthesis takes place on the ribosomes of cells. This section provides a brief introduction to the make-up of nucleic acids, carriers of genetic information.

Nucleic acids, whether DNA or RNA, are polymers composed of *nucleotide* monomers. Each nucleotide contains a five-carbon sugar to which a phosphate group and an organic base are attached (Figure 2.23). In DNA the five-carbon sugar is deoxyribose **2.23**. (Can you recognize that this is a sugar?) The organic bases belong either to the group of pyrimidine bases with one ring, or to the group of purine bases with two rings. All have nitrogen atoms in their ring structures. The four organic bases in DNA are cytosine (abbreviated as C) **2.24** and thymine (T) **2.25** with pyrimidine rings, and adenine (A) **2.26** and guanine (G) **2.27** with purine rings. Nucleotides contain one of these bases. A nucleotide **2.28**, with adenine as its base, is shown in Figure 2.23.

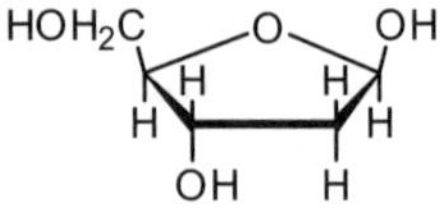

Figure 2.23 Components of DNA nucleotides.
DNA contains deoxyribose **2.23** as the five-carbon sugar, the pyrimidine bases cytosine **2.24** and thymine **2.25**, and the purine bases adenine **2.26** and guanine **2.27**. Structure **2.28** shows a nucleotide.

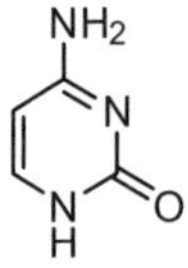

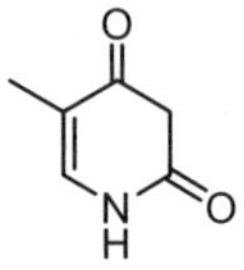

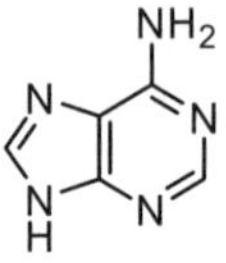

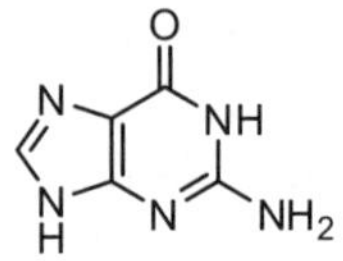

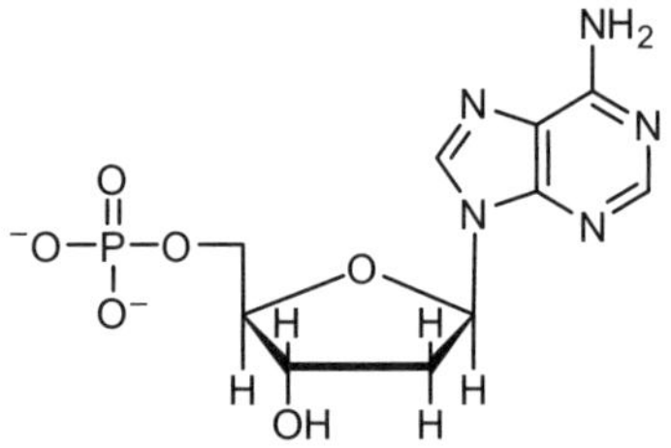

The nucleotides, with their different bases, compose a strand of the polymeric nucleic acid. DNA molecules are usually double-stranded and form the famous double helix, with the sugar-phosphate groups of the nucleotides forming the backbone (Figure 2.24). The pyrimidine or purine bases are oriented towards the interior of the helix. Hydrogen bonds link the bases of the two strands. But only specific bases can pair up, due to their relative sizes and shapes. The possible base-pairs are adenine (A) and thymine (T), or cytosine (C) and guanine (G), as shown in Figure 2.24. DNA molecules continuously reproduce themselves during cell division, and the information is copied to new cells. Sequences of three base pairs code for specific amino acids, according to the genetic code. Messenger RNA reads the instructions from DNA and transports them to the ribosomes where the information is translated into specific amino acids.

The Molecular Building Blocks 73

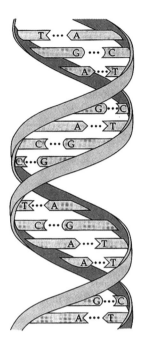

Figure 2.24 The double helix.
(Image from Neil A. Campbell, Biology, 4th edn, © 1996, p. 85. Reprinted by permission of Pearson Education, Inc., Upper Saddle River, NJ.)

Genes can be defined as segments of DNA, *i.e.* their base-pair sequences, that code for a type of protein or RNA that have a function in the organism. The *genome* describes the total of all the genetic information on the DNA of an organism. Different organisms have different genomes; even closely related individuals have slight differences in the make-up of their base sequences.

We have earlier encountered a structure that is related to nucleotides: the common energy carrier adenosine triphosphate (ATP) **2.29** (Figure 2.25). Do you recognize its purine base component? Note that ATP has three phosphate groups. The bonds between these phosphate groups can be broken easily by *hydrolysis* (*i.e.* when reacting with water), in an exothermic, heat-releasing process. This energy-providing reaction forms adenosine diphosphate (ADP), with only two phosphate groups, or adenosine monophosphate (AMP), with one phosphate group left. The released phosphate groups from ATP, in interaction with helpful

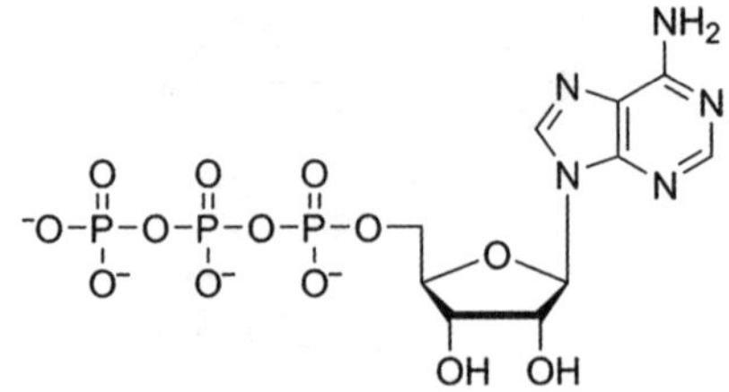

Figure 2.25 Structure of ATP, principal energy carrier.

enzymes, can be attached to other molecules. This process, called *phosphorylation*, activates the recipient molecules for further reactions in organisms.

2.7 REACTION PATHWAYS THAT LINK PLANT MOLECULES

Although plants produce a wealth of different organic compounds, with very diverse chemical structures, they follow some general reaction pathways to assemble them.[15–17] A simple scheme of some of these pathways is shown in Figure 2.26.

At the origin of all plant reactions are photosynthesis and the production of glucose. Simple sugars can be transformed into carbohydrate polymers, like starch and cellulose. Or, sugars are broken down during respiration, providing energy. Other reaction steps lead to the composition of key compounds, like phosphoenolpyruvate **2.30**, isopentenyl pyrophosphate **2.31**, or acetyl coenzyme A **2.32** (Figure 2.27). Plants make use of these key metabolites to synthesize very diverse compounds in further reactions. A study of the reaction pathways, including their key compounds, can provide understanding of structural features in plant molecules. For example, acetyl coenzyme A (or acetyl CoA) can release acetyl groups (highlighted in Figure 2.27), with two carbon atoms. Acetyl CoA is repeatedly involved in the biochemical assembly of fatty acids in plants. As a consequence, fatty acid groups in plant oils all have even numbers of carbon atoms, like 14, 16, 18, or 20 carbon atoms. (Note also the nucleotide structure and energy-rich phosphate groups in acetyl CoA.)

In later chapters we will reflect on more biochemical connections. A large group of plant compounds called *terpenes* will be introduced in the next chapter in their role as plant odors. These

The Molecular Building Blocks

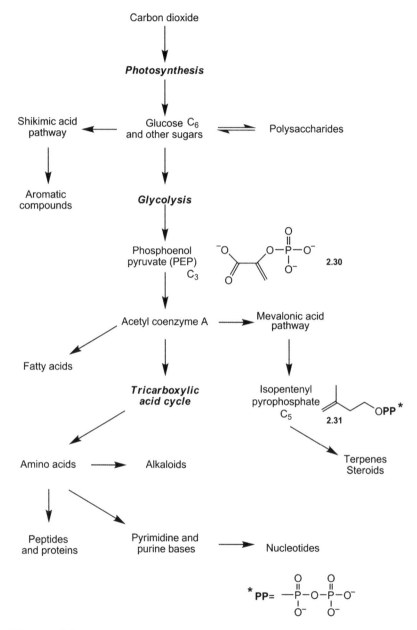

Figure 2.26 Simple scheme of biochemical reaction pathways in plants.

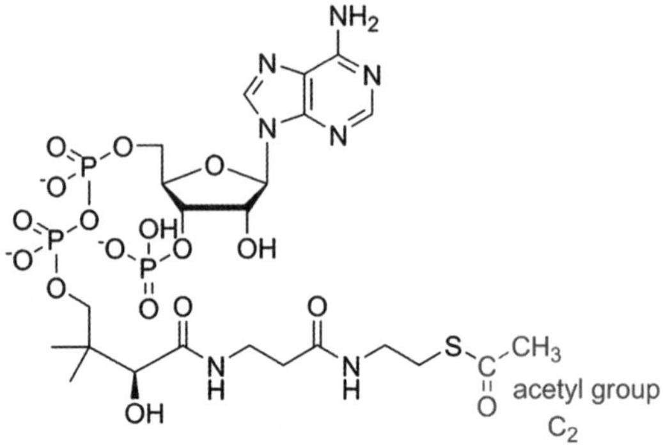

Figure 2.27 Acetyl coenzyme A, a key compound in biosynthetic pathways.

compounds have again typically patterned structures that are the result of following characteristic reaction pathways in plants.

2.8 CONCLUSION

Plants are the source of many of the ubiquitous molecular building blocks that all living organisms need in order to compose the organic compounds in their systems. This chapter introduced the structures of these primary metabolites, along with some important chemistry concepts.

Carbohydrates served as an introduction to asymmetric compounds. Numerous plant (and animal) compounds have molecules with asymmetric structures, often with several chiral centers. Frequently, only the correct orientation of all the asymmetric centers will allow proper functioning in the living systems. We met this type of stereoisomerism also in the structures of amino acids. Only left-handed or L-forms of the amino acids are part of proteins.

Another type of stereoisomers, in the form of *cis* and *trans* isomers, was introduced with fats and oils. Plant oils are usually highly unsaturated, with many double bonds, and have a *cis* arrangement of their double bonds. The rigid double bonds form bends in the hydrocarbon chains that lead to less dense packing of the molecules and with this to lower melting temperatures than in saturated fats. Plant oils are liquids at ambient temperature.

Polymers, huge molecules with monomer units as repeat patterns, were encountered all through the chapter: as starch and cellulose, as lignin in woody plant structures, in highly water-repelling suberin, as the huge and highly diverse protein molecules, and as nucleic acids. Hydrogen bonds within the large molecules of proteins and nucleic acids determine the shapes of these polymers.

Principles of polarity were revisited. Carbohydrates with their numerous OH groups are highly polar, water-soluble compounds because of their many OH functional groups. In contrast, fats and oils, and lipids in general, are hydrophobic compounds due to their large hydrocarbon components. Compounds with both polar and non-polar regions in their molecules were introduced in the form of phospholipids.

Functional groups, attached to the carbon backbone of molecules, are responsible for typical chemical reactions of plant compounds—and of compounds in general. Several types of functional groups were introduced: hydroxy groups (OH) with carbohydrates, ester groups (COOR) with fats and oils, aromatic compounds (with benzene rings) in monomers of lignin, carboxylic acid groups with fatty acids and amino acids, amino groups (NH_2) with amino acids, and peptide bonds (CONH) that link amino acid units into peptides and proteins. Amino acids also recalled acid and base properties.

The genetic information for the assembly of amino acids into proteins is stored on segments of the nucleic acids. We looked at the structures of nucleotides that compose the complex polymeric structures of nucleic acids, in preparation for the later chapter on genetically modified plants.

The chapter concluded by showing some general biochemical reaction pathways that link the different families of plant compounds.

We have now built the chemical foundations to proceed to plant compounds that are derived from the primary metabolites. It is there where plants have evolved an incredible wealth and diversity of substances, namely in the form of plant odors, pigments, and defensive compounds.

REFERENCES

1. H.-D. Belitz, W. Grosch and P. Schieberle, *Food Chemistry*, Springer-Verlag, Berlin, 3rd edn, 2004.
2. P. Le Couteur and J. Burreson, *Napoleon's Buttons: 17 Molecules that Changed the World*, J. P. Tarcher/Penguin, New York, NY, 2004.

3. J. B. Harborne, *Introduction to Ecological Biochemistry*, Academic Press, London, 4th edn, 1993.
4. D. W. Ball, *J. Chem. Educ.*, 2007, **84**, 1643.
5. D. M. Holstege, B. Puschner and T. Le, *J. Agric. Food Chem.*, 2001, **49**, 1648.
6. J. R. Hanson, *Chemistry in the Garden*, The Royal Society of Chemistry, Cambridge, 2007.
7. E. Sjöström, *Wood Chemistry: Fundamentals and Applications*, Academic Press, San Diego, CA, 2nd edn, 1993.
8. N. Gierlinger, L. Sapei and O. Paris, *Planta*, 2008, **227**, 969.
9. B. B. Simpson and M. C. Ogorzaly, *Economic Botany*, McGraw-Hill, New York, NY, 3rd edn, 2001.
10. P. E. Kolattukudy, *Can. J. Bot.*, 1984, **62**, 2918.
11. M. A. Bernards, *Can. J. Bot.*, 2002, **80**, 227.
12. M. Luckey, *Membrane Structural Biology*, Cambridge University Press, New York, NY, 2008.
13. J. B. Reece, L. A. Urry, M. L. Cain and S. A. Wasserman, *Campbell Biology*, Benjamin/Cummings, Menlo Park, CA, 9th edn, 2010.
14. D. L. Nelson and M. M. Cox, *Lehninger Principles of Biochemistry*, W. H. Freeman, New York, NY, 4th edn, 2005.
15. J. R. Hanson, *Chemistry in the Garden*, The Royal Society of Chemistry, Cambridge, 2007.
16. P. B. Kaufman, L. J. Cseke, S. Warber, J. A. Duke and H. L. Brielmann, *Natural Products from Plants*, CRC Press, Boca Raton, FL, 1999.
17. J. McMurry and T. Begley, *The Organic Chemistry of Biological Pathways*, Roberts and Company Publishers, Greenwood Village, CO, 2005.

CHAPTER 3
Perfumes, Fragrant or Foul

3.1 INTRODUCTION

Imagine the smell of a rose on a warm day. The fragrant odors are created by molecules dancing in the air and reaching our noses—or the sensory organs of insects. Many flowers develop distinct fragrances when they open, attracting pollinators like bees or moths. Common plant names like "sweet clover" and "sweet cicely," and trade names like "Fragrant Cloud" for a rose (Figure 3.1(a)) suggest that humans find these odors attractive, too. On the other hand, some flowers smell like decomposing animal material. Common names like "stinkbells" or "corpse flower" (a name that applies to *Amorphophallus titanum* in Figure 1.33 as well as to the giant, foul-smelling flower of *Rafflesia arnoldii* in Figure 3.1(b)) indicate that these flower odors are clearly not attractive to humans. But they are alluring to flies or beetles! Ripening fruits often develop smells as a way of appealing to the animals that feed on them.

Not all plant odors act as attractants for animals, of course: the strong odors found in many leaves, such as in leaves of sage or mint, have defensive functions, keeping animals from feeding on the plants. Here, we will focus on smells of flowers and fruits that are inviting, at least to certain animals.[1,2]

Plant odors are mixtures of many different organic compounds.[3] The fragrance of a rose, for instance, is composed of more than two

---

The Chemistry of Plants: Perfumes, Pigments, and Poisons
Margareta Séquin
© Margareta Séquin 2012
Published by the Royal Society of Chemistry, www.rsc.org

(a)   (b)

Figure 3.1   Perfumes, fragrant or foul.
(a) A rose with the trade name "Fragrant Cloud" gives off a strong, sweet scent. (b) *Rafflesia arnoldii* or "corpse flower", a parasitic plant from the tropical rainforest of Indonesia, develops an intense smell of carrion when its giant flower is in bloom. (Photo by Ch'ien C. Lee.)

hundred different compounds, all combining to create the scent we experience. Sometimes there are a few compounds that stand out with their dominant odors and characterize the entire scent, like the sweet smell of the compound geraniol **3.3** (Figure 3.4) that we recognize as the typical smell of fragrant roses.[4] Plant odors comprise fragrant mixtures of non-polar compounds and therefore do not dissolve in water. (Yet, on a rainy day they can form weak bonds with rain droplets and hitch a ride on the drops, contributing to the intensified plant smells after a rain storm.) The organic molecules that compose plant odors have up to about fifteen carbon atoms. The relatively small size of the odorous molecules (on a molecular scale) lets them easily acquire enough kinetic energy on a warm day to evaporate into the air. Another name for these fragrant oily mixtures is *essential oils*, an expression that is derived here from the word "essence." Indeed, many essential oils from plants provide the fragrances, or essences, that we use in perfumes and other cosmetics, as will be addressed in the later chapter on Plants and People.

Plant smells can be grouped in families of chemically related compounds. Many pleasant fragrances are composed of *terpenes*, with molecules having ten or fifteen carbon atoms arranged in a typical pattern (Figure 3.4). Low-molecular-weight aromatic compounds, with a benzene ring, include the popular smell of vanillin **3.5** (Figure 3.6). Functional groups attached to the carbon backbone of the molecules can characterize or enhance smells. Relatively small molecules with an ester functional group add a pleasing odor to various ripe fruits (Figure 3.7). On the other hand, amines and sulfurous compounds produce many of the foul odors that remind us of rotten meat or decaying fish (Figure 3.8).

The organic compounds that compose plant odors do not occur in every plant, but only in certain species. We can often match the odors with specific plants that emit them: the sweet smell of vanillin reminds us of orchids, or the smell of limonene **3.4** recalls lemons and citrus flowers. Similarly, plant pigments and plant toxins are found in certain plants only. This is in contrast to sugars, fats, and amino acids that are universally found in plants as the primary metabolites. Plant smells, pigments, and plant defenses are sometimes called *secondary metabolites*. Formerly thought to be merely waste products, many secondary metabolites have vital functions in plants, like attracting pollinators. Plants assemble these compounds from the primary metabolites, following distinct biochemical reaction pathways, as shown in Figure 2.26.

3.2 FRAGRANT TERPENES

A large group of plant smells with pleasant odors (again from a human point-of-view) belongs to the family of *terpenes*.[5] If we smell a rose, or flowers of sweet peas (*Lathyrus odoratus*), or blossoms of citrus trees (Figure 3.2), the pleasant odors that we experience are dominated by terpenes that are part of complex mixtures of fragrant compounds. The name "terpene" is derived from "oil of turpentine," itself a mixture of terpenes, with a well-known typical odor. (Read more about oil of turpentine in Chapter 5 on plant defenses.) Terpene molecules are largely hydrocarbons, composed mainly of hydrogen and carbon atoms. This makes them non-polar. Like other hydrocarbon mixtures—think crude oil or oil of turpentine—they do not dissolve in water; they float on it. Many pleasant plant odors are composed of terpenes with ten carbon atoms. At ambient temperatures they are oily liquids that evaporate easily into the air on a warm day.

The most striking feature of terpenes is that they are all composed of five-carbon puzzle pieces, called *isoprene units,* **3.1**. Figure 3.3 shows an isoprene unit as a line structure and as a "ball-and stick" figure. The name comes from the compound isoprene

Figure 3.2 Flowers with fragrant terpenes.
(a) Flowering sweet peas (*Lathyrus odoratus*). (b) Blossoms of a lemon tree (*Citrus* spp.).

Figure 3.3 The isoprene unit in terpenes.
The five-carbon isoprene unit **3.1** is both shown as a line structure and as a "ball-and-stick" figure. The compound isoprene **3.2** is shown on the right.

3.2, and another name for terpenes is *isoprenoids*. Each isoprene unit has four carbon atoms connected to each other in sequence and a one-carbon side branch attached to one of the middle carbons. The more branched end of the puzzle piece is the "head" and the other end the "tail" of the unit.

Now examine the molecule of geraniol **3.3** in Figure 3.4, and also of limonene **3.4**, the fragrant terpene that is found in flowers, fruits, and leaves of citrus trees. Geraniol and limonene are both shown as line structures and as ball-and-stick figures. Note that the carbon skeleton of both molecules is composed of two isoprene units, highlighted in the line structures, and that the units are connected "head" to "tail." (Disregard any functional groups or double bonds.) The pattern of isoprene units, like building blocks, can be detected in every terpene. This means that the total number of carbon atoms in a terpene molecule is always a multiple of five. The isoprene units can be connected "head-to-tail", or "tail-to-tail", or even "head-to-head". Terpenes with ten carbon atoms are called *monoterpenes*; geraniol and limonene are two examples. If a molecule has a total of ten, fifteen, twenty, or forty carbons (or other multiples of five), check for the isoprene puzzle pieces. If you find them, you are looking at a terpene structure. In addition, functional groups may be bonded to the hydrocarbon backbone: in geraniol there is an alcohol (OH) group attached. Such functional groups add special odor notes.

Many terpene molecules contain chiral centers and can therefore occur in plants as different stereoisomers. The molecule of limonene has one center of asymmetry, identified by an asterisk in the molecular structure (Figure 3.4, **3.4**). Interestingly, the two mirror images of limonene smell different to us. One provides the smell of lemon, the other the smell of orange. While the two smells are related, humans can clearly distinguish between them.

Fragrant plant oils have ever fascinated humans, and chemists studied their composition in detail in the 19th century. As early as 1887, O. Wallach proposed that the molecules of many fragrant plant compounds, including geraniol and limonene, are hypothetically constructed from isoprene units. He obtained a Nobel prize

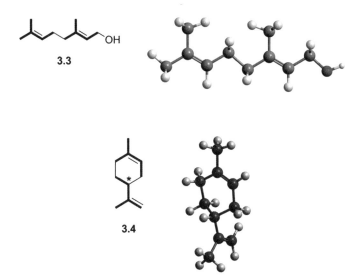

Figure 3.4   Structures of fragrant terpenes.
Geraniol, **3.3** with the molecular formula $C_{10}H_{18}O$, and limonene, **3.4** $C_{10}H_{16}$, are both monoterpenes. Each compound is shown as a line structure, with the isoprene units highlighted, and as a ball-and-stick figure. (Only one of the enantiomers of limonene is shown.)

for his work in 1910.[6] In 1939, L. Ruzicka was awarded another Nobel prize linked to the elucidation of the isoprenoid structure of terpenes.[7] In his extensive research, he showed that plants follow distinct reaction pathways to assemble terpenes: energetically rich five-carbon units, derived from isopentenyl pyrophosphate (see Figure 2.26 in the previous chapter), combine to increase the number of carbons in multiples of five, following the so-called "isoprene rule".

Terpenes are widespread in the plant kingdom. They not only produce attractive fragrances in flowers and fruits. They also provide strong defensive odors in leaves (think of crushed leaves of eucalyptus or mint), pigments in the form of orange and red carotenoid pigments (with forty carbon atoms in their molecules), polymeric rubber particles in milky plant saps, plant hormones, and even plant poisons. We will encounter them again in the later chapters. At this point check the molecule of chlorophyll *a*, **1.1** shown in Chapter 1, and examine its long hydrocarbon tail (the phytyl group). Can you detect the isoprene units there? Evidently plants have evolved successful pathways to link isoprene units and produce numerous vital substances with diverse functions. So, look out for the typical five-carbon patterns in plant molecules.

3.3 SWEET VANILLA AND OTHER AROMATICS

Not all odors that are pleasant to humans are terpenes. Equally alluring are some of the aromatic compounds—which here means they have a benzene ring (Figure 3.6). The molecules of sweet-smelling vanillin **3.5**, a popular flavoring worldwide, have such aromatic rings. Many orchids, like the small alpine orchid shown in Figure 3.5(a), produce vanillin in their blossoms.[8] The vanilla orchid (*Vanilla planifolia*, Figure 3.5(b)) produces it in its seed pods, the vanilla beans.[9,10] (Actually, vanillin in the pods is bonded to glucose and has no odor; only a curing process of the vanilla beans brings out vanillin.) Another fragrant aromatic compound is eugenol **3.6**; it is part of the smell of carnations (Figure 3.5(c)). We know its pleasant smell as the typical odor of cloves, a spice that consists of the dried flower buds of the tropical bush *Syzygium aromaticum*. Both eugenol and vanillin happen to be aromatic in more than one sense: they have a benzene ring in their structure and give off a pleasant odor. The functional groups attached to the

Figure 3.5 Plants with fragrant aromatic compounds.
(a) The small alpine orchid *Nigritella rubra* has an intense smell of vanilla. (b) A vanilla orchid plant (*Vanilla planifolia*) and cured vanilla beans. (c) Carnation flower (*Dianthus caryophyllus*).

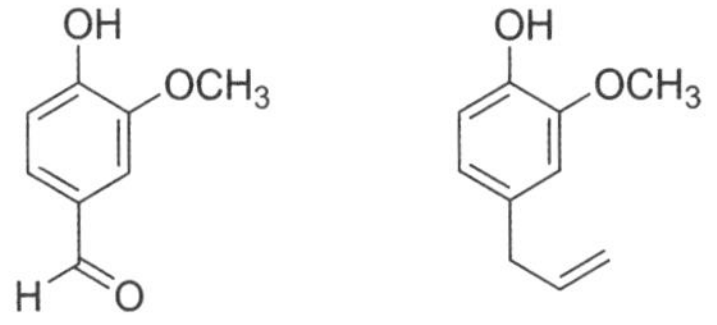

Figure 3.6 Structures of fragrant aromatic plant compounds.
Vanillin **3.5** is found in many orchids; its glycoside occurs in the seed pods of vanilla orchids. Eugenol **3.6** provides the typical smell of cloves and is also part of the fragrance of carnations (*Dianthus* spp.).

benzene rings of vanillin and eugenol contribute to the pleasing character of the odors.

3.4 PLEASANT-SMELLING ESTERS

Many fruits are meant to be eaten by animals that will then digest the food and spread the seeds in their feces—which makes good fertilizer for the sprouting seeds. The texture, taste, smell, and color of fruits change during the ripening process. Sour or otherwise distasteful compounds that protected the unripe seeds are transformed into tasty sugars and attractive odors. As humans, we are quite familiar with inviting odors of fruit selected for its ripeness. We recognize the odor of a ripe apple or banana. Both smells are dominated by fragrant esters (we encountered

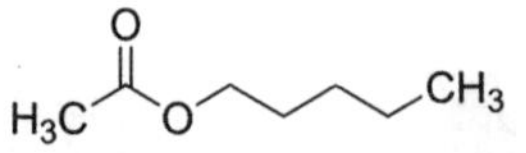

Figure 3.7 Structures of pleasant-smelling esters.
The ester amyl acetate **3.7** characterizes the smell of ripe bananas. Ethyl 2-methylbutyrate **3.8** provides the dominant odor of ripe apples.

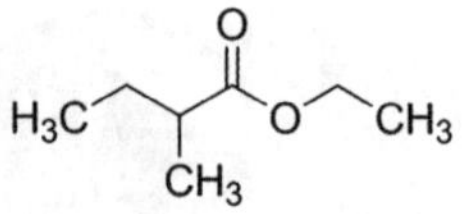

this functional group in fats and oils).[11] The pleasant odors of ripe fruits are usually produced by mixtures of volatile esters (Figure 3.7) or compounds with aldehyde (CHO) or alcohol (OH) functional groups. Again, often one odor note, provided by a specific compound, stands out in the fragrance of a particular fruit. An ester with the name of amyl acetate **3.7** provides the typical smell of bananas. Ripe apples are recognizable from the smell of the ester ethyl 2-methylbutyrate **3.8**.

3.5 MALODOROUS AMINES AND SULFUROUS COMPOUNDS

Finally, we consider foul smells—the malodorous mixtures that are attractive to flies and sometimes beetles, too. When the giant *Amorphophallus titanum* is blooming (Figure 1.33), flies are buzzing in swarms above its inflorescence. Flowers or clusters of flowers that give off these foul odors often have a maroon color. (See also the giant *Rafflesia* flower in Figure 3.1(b)). The odors and the color may be imitating decomposing animal material.

Foul and fetid plant smells are usually created by sulfurous compounds; some amines, with a -NH or $-NH_2$ functional group, can contribute to them, too (Figure 3.8).[12,13] Their molecules have low molecular weights and evaporate easily. Many foul-smelling compounds have the odor of carrion and attract flies or beetles that pollinate the plants. The names of some compounds, like putrescin **3.9** or cadaverine **3.10**, describe their odors vividly. Some off-odors from plants are only mildly unpleasant to humans. Beetle pollinators are lured by the off-smell of certain magnolia blossoms (Figure 3.9(a)), for instance. The odor is dominated by some simple-structured hydrocarbons, like pentadecane with

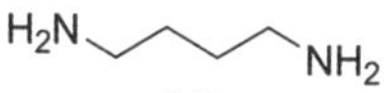

$H_2N\diagup\diagdown\diagup NH_2$ $H_2N\diagup\diagdown\diagup\diagdown NH_2$
   3.9    **3.10**

3.11

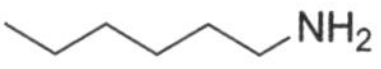

$\diagup\diagdown\diagup\diagdown NH_2$
3.12

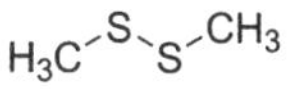

$H_3C-S-S-CH_3$ 3,5-dimethyl-1,2,4-trithiolane
    3.13 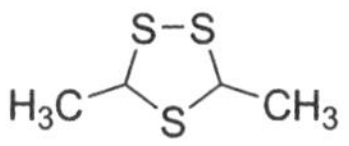   **3.14**

Figure 3.8 Structures of malodorous plant compounds.
Putrescin **3.9** and cadaverine **3.10** are two diamines, each having two NH_2 groups, with highly unpleasant smells. Pentadecane, **3.11** $C_{15}H_{32}$, with a mild off-smell, attracts beetle pollinators to magnolia blossoms. 1-Aminohexane **3.12**, found in the flowers of cow parsnip (*Heracleum lanatum*), attracts flies. Dimethyl sulfide **3.13** has a foul odor and is part of the smell of "corpse flower". The sulfurous molecule **3.14** (3,5-dimethyl-1,2,4-trithiolane) is the strongest (mal)odorant in the durian fruit.

(a)   (b)

Figure 3.9 Flowers with mild off-odors.
(a) Magnolia blossom (*Magnolia* cultivar). (b) Cow parsnip (*Heracleum lanatum*).

fifteen carbon atoms, **3.11**. A simple amine, 1-aminohexane **3.12**, produces the faint carrion smell of cow parsnip (*Heracleum lanatum*, Figure 3.9(b)) and attracts flies for pollination. The full odors of these flowers, as always, are composed of mixtures of volatile molecules.

Figure 3.10 Sulfurous smells in plants.
(a) Durian fruit (*Durio* sp.) and (b) durian fruit cut open, showing a fleshy mass that has a penetrant odor.

As for fruits with a penetrating off-smell, the most famous one must be durian (Figure 3.10), a tropical fruit from the durian tree (*Durio* spp.). Some people regard its edible flesh a delicacy; but the fruit is expressly forbidden in many hotels because of its overpowering smell. Yet, its strong odor, dominated by sulfurous compounds like dimethyl disulfide **3.13** and the malodorous, sulfurous compound **3.14**, proves irresistible to the animals that feed on the fruit and in the process spread its large seeds.[14]

3.6 ANALYZING PLANT ODORS

How can the components of a plant smell be identified, especially when considering that plant scents are often intricate mixtures of many different compounds? Traditionally, fragrant plant materials have been treated with steam, in so-called steam-distillations. In this method, steam carries the volatile oils out of a plant sample, and the distillate is collected. The layer of essential oils, floating on water, can then be separated. Still a complex mixture of many compounds, the fragrant oily layer can subsequently be further separated and analyzed.[15,16]

Often, only very small amounts of fragrant materials are available, as from the odors of flowers and fruits. A steam-distillation would be impossible to use there. A method, called "head-space analysis", can be used instead in which special glass bell jars, containing materials that absorb odors, are placed over the fragrant plant parts. This method has made it possible to trap and collect traces of volatile oils.[17]

Modern instruments can separate many of these fragrant mixtures and can identify their components, of which there are sometimes hundreds from a single fragrant plant. A very sensitive method that is commonly used involves the combination of gas chromatography (GC) and mass spectrometry (MS). This method is generally useful for defining the volatile components in a mixture, as in a plant oil. (GC/MS also has vast applications in analyzing mixtures like food flavorings, artificial perfumes, or samples of petroleum.)

A gas chromatograph is an instrument that can separate very small amounts (like a tiny drop) of volatile mixtures. It is basically an oven, with controllable temperature, in which a very thin capillary tube is suspended (Figure 3.11(a)). The inside of this capillary tube is coated with an adsorbent. After a sample of a volatile oil is injected onto the column, a stream of gas, like helium gas, carries the mixture through the capillary. During this process, the components of the mixture separate according to the sizes of the different molecules, their polarities, and their interactions with the adsorbent. While small, non-polar molecules travel fastest through the capillary and then exit, larger less volatile molecules pass more slowly through the system. The exiting compounds are recorded by a detector. Each peak on the resulting gas chromatogram represents a different compound. Figure 3.11(b) shows a gas chromatogram of rose oil, the fragrant mixture that is obtained by extracting the petals of roses. Note the numerous peaks in the gas chromatogram, each of them the result of a different compound (one of them being geraniol). While a gas chromatograph can separate the mixtures, it does not identify the components. Frequently, another instrument called a mass spectrometer is connected to the gas chromatograph which further evaluates each peak (*i.e.* each compound). A mass spectrometer provides information about the mass and the composition of molecules and thus can determine the structures of the compounds.

Another method to evaluate the results from gas chromatography uses a "sniffing port" attached to the exit of the instrument; traces of each volatile compound separated out can be sniffed and evaluated. Quite amazingly, very different odors, from sweet to

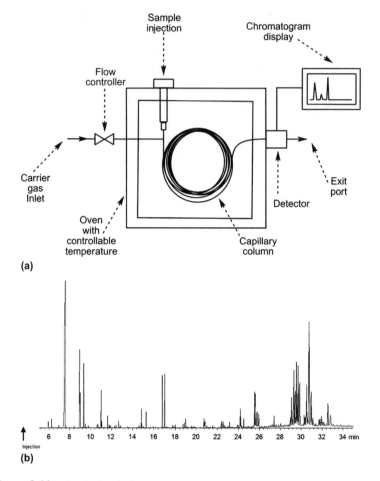

Figure 3.11   Analysis of plant scents.
(a) Schematic drawing of a gas chromatograph. (b) Gas chromatogram of rose oil. Each line in the chromatogram represents a different compound. (Gas chromatogram by Phenomenex, Inc.)

earthy to downright malodorous, can contribute to one particular plant scent.

3.7   CONCLUSION

In this chapter we examined odorous compounds from plants that serve as attractants in flowers and fruits. There are some common features to the composition of plant smells: they are complex mixtures, composed of many non-polar organic compounds. Their molecules are of relatively small size (on a molecular scale), and have few, if any, polar functional groups like OH bonded to their carbon structures. This means that there can only be weak attractions between the molecules. As a consequence, the odorous compounds rapidly evaporate when temperatures rise. Gas chromatography, combined with mass spectrometry, is a highly sensitive method that can detect and characterize components in very small samples of volatile mixtures, like those of a plant smell.

Some plant odors are generally experienced as pleasant fragrances by humans. These odors have certain common chemical characteristics. They often contain compounds with the typical isoprene pattern of terpenes, or with benzene rings, or with functional groups like esters or aldehydes. They are attractive also to many animals that act as pollinators. Amines and sulfurous molecules, on the other hand, provide some foul plant odors that are nevertheless appealing to flies and beetles.

Many molecules of smelly plant compounds have asymmetries in their structures. This means that different stereoisomers are possible. Interestingly, they often have distinctly different smells.

While many odorous plant compounds serve as attractants for animals in fruits or in flowers, others act as deterrents, to keep animals from eating the plants. The strong, deterring odors are usually found in leaves, and some examples will be addressed in the later chapter on plant defenses.

REFERENCES

1. J. B. Harborne, *Introduction to Ecological Biochemistry*, Academic Press, London, 4th edn, 1993.
2. J. R. Hanson, *Chemistry in the Garden*, The Royal Society of Chemistry, Cambridge, 2007.
3. R. Teranishi and S. Kint, in *Bioactive Volatile Compounds from Plants*, ed. R. Teranishi, R. G. Buttery and H. Sugisawa, ACS Symposium Series, American Chemical Society, Washington, DC, 1993, **525**, ch. 1, pp. 1–5.
4. A. Antonelli, C. Fabbri, M. E. Giorgioni and R. Bazzocchi, *J. Agric. Food Chem.*, 1997, **45**, 4435.

5. E. Breitmaier, *Terpenes: Flavors, Fragrances, Pharmaca, Pheromones*, Wiley-VCH Verlag, Weinheim, 2006.
6. M. Christmann and Otto Wallach, *Angew. Chem., Int. Ed.*, 2010, **49**, 9580.
7. L. Ruzicka, *Proc. Chem. Soc.*, 1959, 341.
8. R. A. J. Kaiser, in *Bioactive Volatile Compounds from Plants*, ed. R. Teranishi, R. G. Buttery and H. Sugisawa, ACS Symposium Series, American Chemical Society, Washington, DC, 1993, **525**, ch. 18, pp. 240–268.
9. B. B. Simpson and M. C. Ogorzaly, *Economic Botany*, McGraw-Hill, New York, NY, 3rd edn, 2001.
10. D. Havkin-Frenkel and R. Dorn, in *Spices: Flavor Chemistry and Antioxidant Properties*, ed. S. J. Risch and C.-T. Ho, ACS Symposium Series, American Chemical Society, Washington, DC, 1997, **660**, ch. 4, pp. 29–40.
11. H.-D. Belitz, W. Grosch and P. Schieberle, *Food Chemistry*, Springer-Verlag, Berlin, 3rd edn, 2004.
12. J. B. Harborne, *Introduction to Ecological Biochemistry*, Academic Press, London, 4th edn, 1993.
13. G. C. Kite and W. L. A. Hetterschieid, *Phytochemistry*, 1997, **46**, 71.
14. H. Weenen, W. E. Koolhaas and A. Apriyantono, *J. Agric. Food Chem.*, 1996, **44**, 3291.
15. J. B. Harborne, *Phytochemical Methods*, Chapman and Hall, London, 3rd edn, 1998.
16. R. Ikan, *Natural Products, A Laboratory Guide*, Academic Press, London, 2nd edn, 1991.
17. R. Kaiser, in *Perfumes, Art, Science, and Technology,* P. M. Muller and D. Lamparsky (ed.), Elsevier, Amsterdam, 1991, 213.

CHAPTER 4
Colorful Plant Pigments

4.1 INTRODUCTION

The palette of plant colors is vast: just think of yellow sunflowers, blue, purple, and green grapes, and red and yellow leaves in fall. Or imagine the many shades of green in leaves, or the varied tones of brown in tree bark (Figure 4.1(a)). Colorful pigments can be found in all parts of plants, including their roots (like carrots or red radishes). The colors are noted by animals—and humans. When strawberries or tomatoes have a deep red color, we are likely to select and eat them. In horticulture, plants are chosen and bred for the pigments in their blossoms, their leaves, and even their roots and tubers (as in different varieties of carrots and potatoes). Plants evolved many of the pigments in order to live and reproduce. Green chlorophyll captures energy from sunlight. Attractive colors of blossoms invite pollinators (Figure 4.1(b)). Bright colors in fruits advertise their ripeness to animals, at the time when seeds are ripe (Figure 4.1(c)). And some pigments protect plants by trapping harmful radiation.

In this chapter, we learn about the major families of compounds that form plant pigments. We'll also examine what parts of their molecules interact with light in a way that makes us see a color. Just as molecules of plant smells have certain common characteristics, so do molecules of plant pigments. The colorful materials in

Figure 4.1 Colorful pigments in plants.
(a) Foliage and bark of blue-gum eucalyptus (*Eucalyptus globulus*).
(b) Bright blossom of *Trichocereus grandiflorus* cactus. (c) Ripe rosehips of *Rosa californica*.

flowers, leaves, and roots are composed of organic compounds, but with generally larger molecules than we encountered in the previous chapter on plant smells. Long sequences of single bonds alternating with double bonds, *i.e.* conjugated double bonds, are characteristic of pigment molecules. You may recall from Section 1.6.3 that these structures have electrons that easily interact with light and, in the process, absorb parts of visible sunlight (Figure 1.31). Wavelengths that are not absorbed by pigments are reflected to us, and their combination appears as the color of a plant part.

Bonds between metal ions and organic molecules also cause strong absorption of light, often within the human visible range. We already encountered chlorophyll *a* and its organic molecular structure that holds a central magnesium ion (Figure 1.29). Metal ions from the soil, like aluminum or iron, can bond with OH functional groups in pigment families like the anthocyanins or the tannins. This creates many different shades of colors, sometimes within the same plant.

Plant pigments can be grouped in a few chemical families that are structurally related to each other.[1-3] Pigments that are chemically similar have common properties, *e.g.* in terms of stability or water-solubility. We will begin with a comparison of different *chlorophylls*, and then go on to *carotenoids* that include many yellow, orange, and red pigments. *Flavones* provide white and yellow colors in plants. They are part of an even larger family of compounds, known as *flavonoids*. Among flavonoids, we will also encounter blue and purple *anthocyanins* and brown *tannins*.

We will then examine a smaller group of purple pigments, unrelated to flavonoids, called *betalains*. They are alkaloids and take the place of anthocyanins in a select group of related plants. In each family of pigments, we will learn about their typical structures, their special properties, and their distribution in plants.

Whether plant pigments are water-soluble or not, determines where in the cells they are stored. Pigments that dissolve in water are stored in the vacuoles in plant cells (see Figure 1.26). Anthocyanins and betalains, and also tannins and flavones, have numerous OH groups in their molecules that can easily form hydrogen bonds with water. In addition, some of their OH groups are bonded to sugars, making them glycosides. Therefore, these pigments are soluble in water. In contrast, pigments like chlorophyll and carotenes that have no (or very few) OH groups do not dissolve in water, but are fat-soluble. They are stored in their own cell organelles: chlorophyll in chloroplasts and carotenes in chromoplasts.

The colors of plants that we see are the result of mixtures of pigments. One of the pigments may be dominant, like chlorophyll in green leaves. Structural colors, caused by plant structures like a rough surface or fine hairs on leaves, can contribute to the color shade of a plant part. But they are not as dominant in plants as, for example, in bird feathers or in wings of butterflies where structural colors create many of the colors that we see there.[4]

For further illustration of plant pigments, Box 4.1 takes a combined look at plant scents and pigments as attractants of pollinators, whereas Box 4.2 reflects on the pigments in fall coloration.

Let us explore now the different families of plant pigments and learn about their special characteristics.

4.2 THE CHLOROPHYLLS

In Chapter 1, we encountered chlorophyll *a*, the main photosynthetic pigment (Figure 1.29, **1.1**). There are other types of chlorophylls that distinguish themselves by slight variations in their molecular structures (Figure 4.2). All chlorophylls have porphyrin rings, with long sequences of conjugated double bonds, and with a magnesium atom in the center of the ring. The differences are in some of the side chains attached to the porphyrin ring. (R and R' in Figure 4.2 point out the sites where the differences are.)

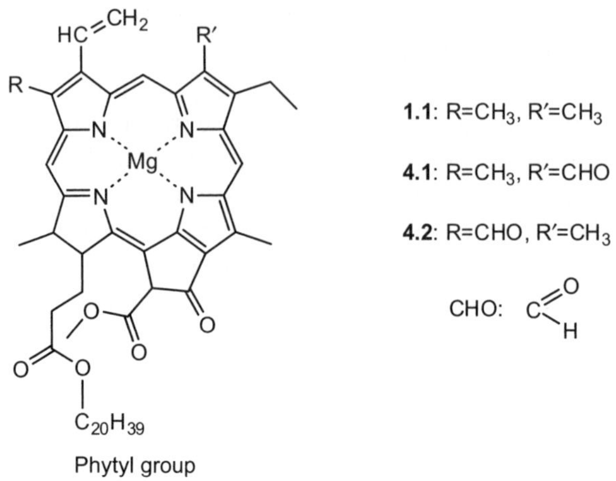

Figure 4.2 Structures of different chlorophylls.
Chlorophyll *a* **1.1**, chlorophyll *b* **4.1**, and chlorophyll *f* **4.2**.

These variations lead to the absorption of different wavelengths of light—and to a slightly different color of the pigment. A side chain that adds an additional double bond in the conjugated bond pattern enhances absorption of light towards longer wavelengths. Note that chlorophyll *b* (Figure 4.2, **4.1**) has an aldehyde (or CHO) group replacing one of the methyl (CH$_3$) groups. The aldehyde group, with a double bond between the carbon and oxygen, is written out for clarity in the figure. If you check the absorption spectra in Figure 1.31, you can compare the wavelengths of maximum absorption for chlorophyll *a* and *b*. Note that in chlorophyll *b* they are shifted towards longer wavelengths. Chlorophyll *b* serves as an accessory pigment in green plants as it broadens the range of light that is available for photosynthesis. Other chlorophylls and related pigments are found in algae or in cyanobacteria. Pigments of photosynthesizing organisms that live in water need to be able to make use of light energy that is available in their environment. Quite recently, researchers discovered a new type of chlorophyll in cyanobacteria from stromatolites (ancient layered calcium carbonate deposits), off the coast of Western Australia.[5] Aside from other wavelengths, this pigment, named chlorophyll *f*, **4.2** absorbs near-infrared. This discovery points to photosynthetic organisms that can make use of longer, low-energy wavelengths in otherwise light-restricted environments.

The name "chlorophyll" comes from the Greek for "green" and "leaf". Chlorophyll molecules contain many carbon atoms and hardly any polar groups, and thus are hydrophobic. The pigments are stored in the chloroplast organelles in plant cells.

Chlorophyll *a* is a fragile compound, sensitive to heat, cold, and strong light. A cold snap in fall will destroy the pigment in green leaves, and so will a heat wave. Strong irradiation from sunlight will lead to pale, sick leaves. (And overcooking green vegetables like spinach will lead to a loss of the deep-green color!) Part of the fragility of chlorophyll is related to the relative ease with which its molecules can lose their central magnesium ion.

It is interesting to note that a similar porphyrin ring structure, with an iron atom at its center, is found in heme, which is part of hemoglobin in red blood cells.

4.3 YELLOW, ORANGE, AND RED CAROTENOIDS

The family of carotenoids includes many yellow and orange plant pigments, and some red ones, too. The name itself is derived from carrots (*Daucus carota*) as the pigment β-carotene **4.3** gives carrot roots their bright-orange color. Carotenoids are wide-spread in plants: they provide orange colors to flowers, like nasturtiums and California poppies (*Eschscholzia californica*, Figure 4.3(a)). They give bright colors to many fruits, like rosehips (Figure 4.1(c)), red peppers, tomatoes (Figure 4.3(b)), and oranges. They are even found in some roots, as mentioned above. Yellow pollen in flowers, so well perceived by bees, gets its color from carotenoids.

(a)　　　　(b)　　　　(c)

Figure 4.3 Carotenoids in flowers, fruits, and leaves.
(a) California poppy (*Eschscholzia californica*). (b) Ripe red tomatoes (*Solanum lycopersicum*). (c) Golden aspen (*Populus tremuloides*) in fall.

In deciduous leaves, carotenoids serve as accessory pigments together with other yellow pigments. They remain disguised by chlorophyll in spring and summer, but show up in fall (Figure 4.3(c)) when chlorophyll decomposes.

Carotenoids are isoprenoid compounds with forty carbon atoms, *i.e.* they are terpenes.[6] If we examine the molecule of β-carotene **4.3** in Figure 4.4, we can discover eight isoprene units linked to each

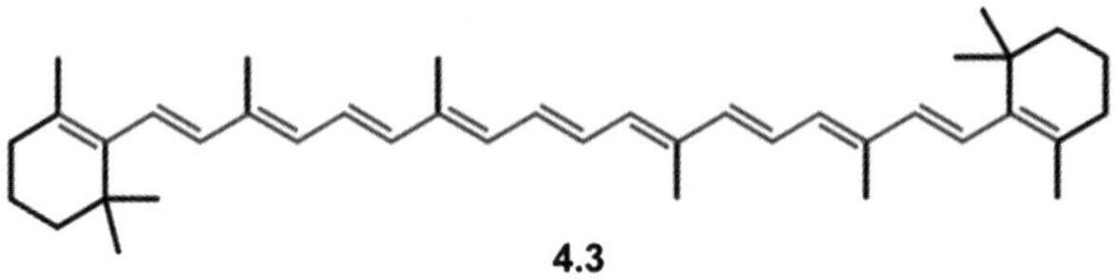

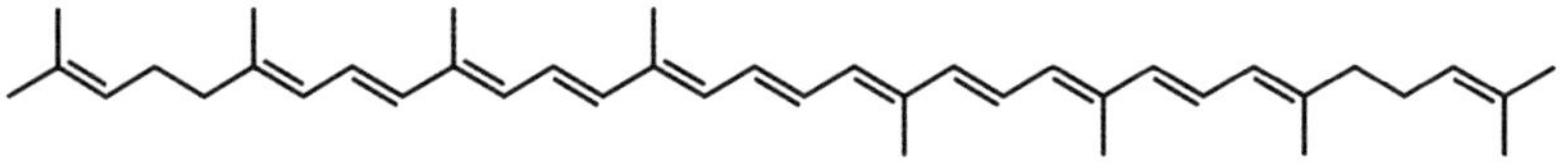

Figure 4.4 Structures of some carotenoids.
β-Carotene **4.3**, $C_{40}H_{56}$, is an orange and yellow pigment in flowers, in deciduous leaves, in many berries and fruits, and in carrot roots. Its long conjugated system is highlighted in red. Lycopene **4.4**, $C_{40}H_{56}$, is the red pigment in tomatoes and rosehips. Phytoene **4.5**, $C_{40}H_{64}$, in unripe tomatoes, absorbs in the UV only. During the ripening process some of the single bonds are oxidized, leading to the longer conjugated sequence in red lycopene. Lutein **4.6**, $C_{40}H_{56}O_2$, is a common yellow xanthophyll in leaves.

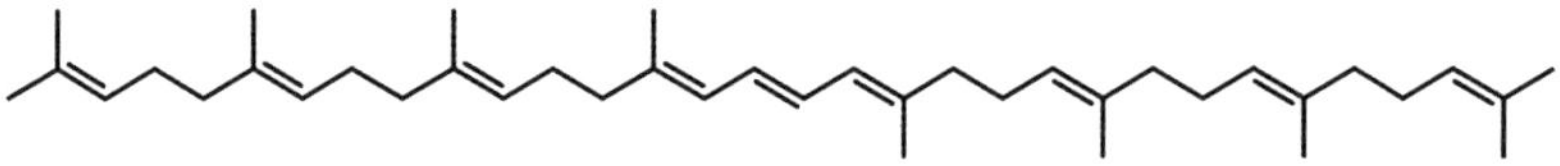

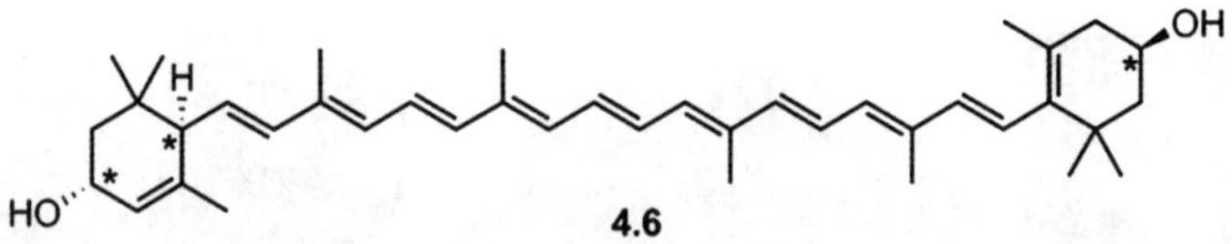

other assembling the pigment. Can you find the units? Notice the long chain of conjugated double bonds (highlighted in red). β-carotene absorbs light in the blue region of the electromagnetic spectrum (Figure 1.31), and, thus, we perceive a yellow-orange colored pigment. Lycopene **4.4** is a red carotenoid, found as the red pigment in ripe tomatoes and rosehips, and in the fleshy part of watermelons. Its structure is slightly different from β-carotene, with additional double bonds. This results in the absorption of different wavelengths, and we consequently see a different color. Unripe tomatoes contain phytoene **4.5**. Note how the pattern of conjugated double bonds is interrupted in this molecule. Its structure absorbs UV-light only, and the pigment appears colorless to us. The green color of unripe tomatoes is provided by chlorophyll. When tomatoes ripen, phytoene is gradually oxidized to red lycopene.

Carotenoids are mostly hydrocarbons. This means they are fat-soluble and do not dissolve in water. (Cooking orange carrots in water will not make them pale!). In plant cells, carotenoids are stored in their own organelles, the chromoplasts. Carotenoids are divided into two subfamilies: the *carotenes* and the *xanthophylls*. While carotenes are true hydrocarbons, consisting of carbon and hydrogen only, the xanthophylls have some functional groups with oxygen, like OH. The name "xanthophyll" is derived from Greek for "yellow" and "leaf", and xanthophylls like lutein **4.6** are yellow pigments in leaves. Lutein (from "luteus", Latin for yellow) is a widespread pigment in deciduous green leaves. It acts there as an accessory pigment. Lutein also strongly absorbs light in the UV region and protects leaves from high-energy radiation that might damage chlorophyll. Xanthophylls are chemically more stable than chlorophyll, and in fall the yellow color of the pigments persists after chlorophyll has decomposed. (Read more about the different pigments in fall coloration and their roles in Box 4.2.) Notice the three chiral centers in lutein, marked by asterisks. A stereoisomer of lutein, zeaxanthin, is another widespread xanthophyll in leaves and provides the yellow color to kernels of corn or maize (*Zea mays*).

Carotenoids have important roles in animal organisms as well: they constitute pigments in egg yolk, in bird feathers, and in the meat of fish like salmon. β-carotene and related compounds can be metabolized in the human body to vitamin A, important for the vision process. But animals and humans cannot synthesize carotenoids; they must obtain them through their plant-related diet.

4.4 WHITE AND PALE-YELLOW FLAVONES

Flavones and closely-related flavonols provide white to pale-yellow colors in flowers and leaves. White daisies and narcissus (Figure 4.5(a)), off-white and yellow chrysanthemums, Cape sorrel (*Oxalis pes-caprae*, Figure 4.5(b)) and white desert poppies (*Argemone* sp., Figure 4.5(c)) have them in their flower petals. Note how strongly the deep-yellow stamens, pigmented by carotenoids, stand out in the desert poppies. Leaves of green deciduous plants contain flavones, hidden by chlorophyll. As flavones strongly absorb light in the UV region, they protect green leaves from damaging radiation together with xanthophylls.

Flavones and flavonols belong to an even larger family of compounds known as *flavonoids*.[7,8] A characteristic three-ring pattern **4.7** is part of all their molecular structures (Figure 4.6). Look for the pattern in the structure of luteolin **4.8**, a colorless to slightly yellow flavone, found, for example in white chrysanthemums. You can find the pattern again in quercetin **4.9**, a common flavonol in many fruits and vegetables. The OH group highlighted in red in Figure 4.6 assigns quercetin as a flavonol.

(a) (b) (c)

Figure 4.5 Flavones in plants.
Flavones and flavonols are white and bright-yellow pigments in the petals of (a) narcissus (*Narcissus* cultivar), (b) Cape sorrel or sourgrass (*Oxalis pes-caprae*), and (c) desert poppy (*Argemone* sp.). They are also in the plants' green leaves. Carotenoids give pollen in (c) its golden yellow color.

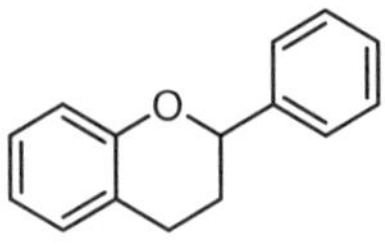

Colorful Plant Pigments

4.7

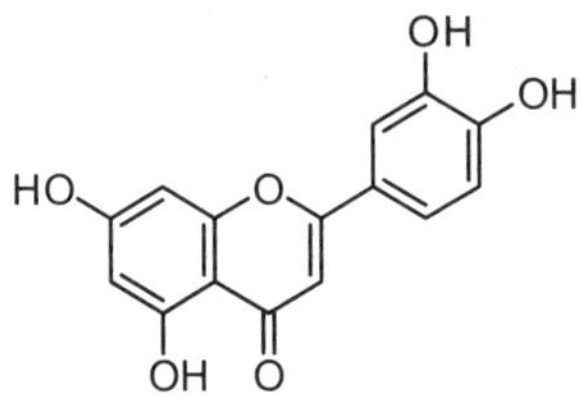

4.8 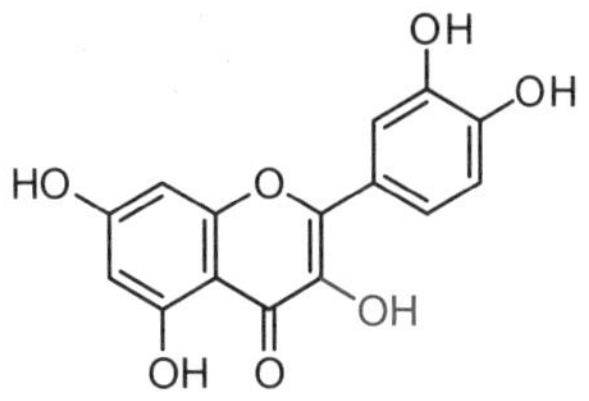 **4.9**

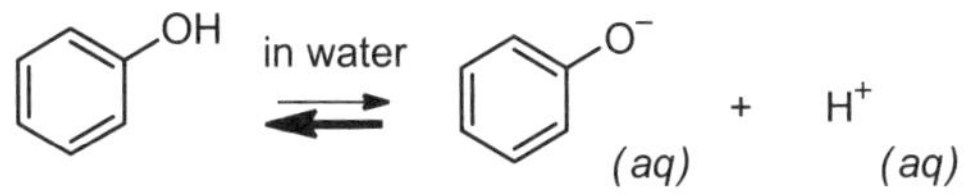

4.10

Figure 4.6 Flavonoid structures.
The characteristic flavonoid ring pattern **4.7** is found in luteolin **4.8**, a flavone, and in quercetin **4.9**, a flavonol. Phenol **4.10** (a compound not found in plants) is a structural group found in all flavonoids. Its property as a weak acid is shown.

Note that the OH groups are directly bonded to aromatic rings. Such an arrangement is called a *phenolic* group. The name is derived from phenol **4.10**, a compound that is not found in nature. Phenolic groups are very weak acids and interact with acids and bases. They can also act as *antioxidants*. This means that they can inactivate oxidizing agents that would damage vital compounds in living organisms. Flavonoids are *phenolics* and have powerful *antioxidant* activity. The later chapter on foods and vitamins from plants will elaborate more on antioxidants.

Flavones and flavonols strongly absorb UV light. While these pigments may appear colorless or barely colored to us, insects like bees detect this type of absorption. Flowers with flavonoid pigments in them are highly visible to bees. White or yellow flowers that may look like one solid color to us show distinct patterns to

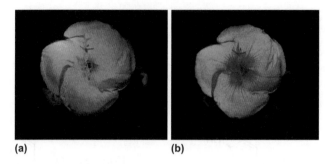

Figure 4.7 Flavones and UV light.
(a) Flower of an evening primrose (*Oenothera* sp.). (b) The same flower viewed under UV-light. (Display at Museu da Ciência, University of Coimbra, Portugal.)

them. UV photography of flowers can give us an idea of the different perception (Figure 4.7). The patterns attract the insects and lead them to nectar and pollen in blossoms. Flavonoid and carotenoid pigments both contribute to the colors of yellow flowers. As the different pigments absorb different wavelengths in the UV, they combine to provide a great variety of patterns that can act as nectar guides for insects, as shown with the example of the flower of an evening primrose (Figure 4.7).

Flavonoids provide most of the bright colors in plants. We will find their typical three-ring pattern again in purple pigments called anthocyanins and in dark-colored tannins. Various numbers of OH groups attached to the framework, together with additional structural differences, lead to great variation in colors. The numerous OH groups make flavonoids at least partially soluble in water. As the pigments are bonded to sugar molecules like glucose, making them glycosides, their water-solubility is further enhanced. The pigments are stored in the cell vacuoles.

4.5 PURPLE, PINK, AND BLUE ANTHOCYANINS

Almost all the purple, reddish-pink and blue colors in plants are created by anthocyanin pigments. The deep-blue color of cornflowers or bachelor buttons (*Centaurea cyanus*) (Figure 4.8(a)), the skin coloration of red apples and blue grapes, and the color of red cabbage leaves are provided by anthocyanins. Roots like red radishes (Figure 4.8(b)) have them in their skins, and red fall leaves

Colorful Plant Pigments

(a)   (b)   (c)

Figure 4.8   Anthocyanin pigments in plants.
(a) Cornflower or bachelor button (*Centaurea cyanus*). (b) Red radish (*Raphanus sativus*). (c) Young leaves of grape vines (*Vitis vinifera*).

get their color from this pigment family. Many young leaves, like the emerging leaves of grape vines (*Vitis vinifera*) (Figure 4.8(c)), are purplish red because of anthocyanins. When plant leaves are injured, they often form reddish pigments along the damaged sections. Stress by cold or drought can bring on red pigmentation by anthocyanins as well. Incidentally, the name "anthocyanins" is related to their color: "Anthos" in Greek means "flower", and "cyanin" relates to "blue". One of the first anthocyanin pigments was isolated from blue cornflowers.

Anthocyanins are a subgroup of the flavonoids. Examine the structures in Figure 4.9 and find the characteristic three-ring pattern **4.7** that we encountered earlier in flavones. Anthocyanins are glycosides. (Without the sugar portion attached they are called anthocyanidins.) They have many OH groups and are stored in the cells' vacuoles. Note that there is an ionic site in the molecules, at the charged oxygen; it enhances the polar character of anthocyanins and promotes their water-solubility. (If you steep red cabbage in hot water, the purple mixture of anthocyanin pigments will color the water.) Pelargonidin **4.11**, as its glycoside called pelargonin **4.12**, provides the various shades of pink, red and purple in pelargonium (and geranium) flowers. It also contributes to the color of ripe raspberries and strawberries. Malvidin **4.13** provides color to blue grapes and to the blue petals of mallow flowers. The long conjugated systems in the molecules lead to strong absorption within the visible range. And let us remember that mixtures of different pigments create the color of a flower, fruit, or leaf.

An interesting specialty of anthocyanins is their color variability—from pink to purple to red to deep-blue. Flowers like blue lupines

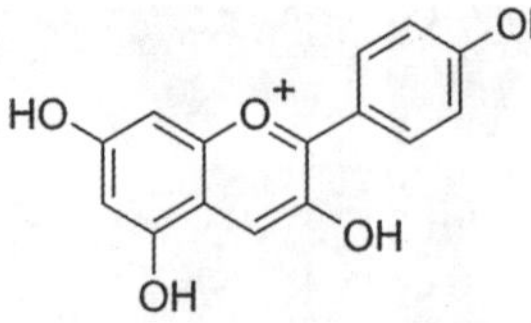

Figure 4.9 Structures of anthocyanidins and anthocyanins.
Pelargonidin **4.11** is a common anthocyanidin. The anthocyanin pelargonin **4.12**, an example of a glycoside, has two glucose groups ("gluc") bonded to its molecules; it is found in pelargonium flowers and ripe raspberries. Malvidin **4.13**, as its glycoside, provides color to purple and blue grapes.

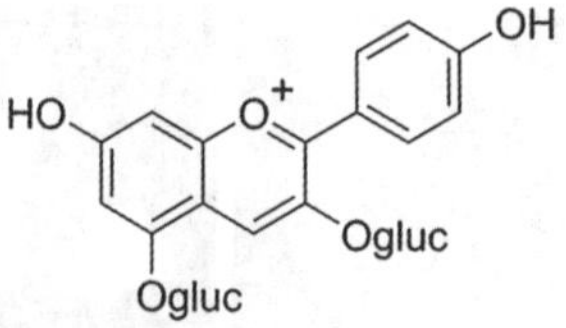

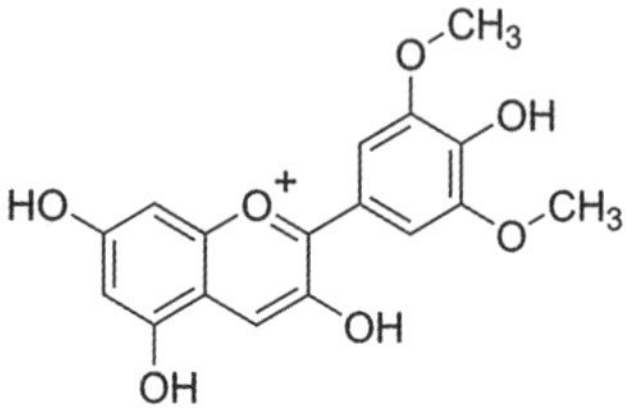

Figure 4.10 Color variation of anthocyanin pigments.
(a) Wild lupine (*Lupinus nanus*) with different shades of anthocyanin pigments. (b) Pink to light blue blossoms of hydrangeas (*Hydrangea* cultivars).

(Figure 4.10(a)) can have quite an array of anthocyanin colors within the same cluster. Several factors create the great palette of colors. Different concentrations of the pigments lead to variations from faint shades to very dark colors, as in black tulips (which actually have a very deep purple color). The difference in the number of

OH groups or OCH_3 groups in the molecules creates many color shades, with additional groups shifting the color to blue (and absorption of light to longer wavelengths). Compare the structures of pelargonidin **4.11** and malvidin **4.13** in Figure 4.9, and note the additional functional groups in the latter.

Colors are further affected by metal ions from the soil that bond to anthocyanin molecules. Hydrangea plants (*Hydrangea macrophylla*) (Figure 4.10(b)), with pink, blue, or purple flowers, are common garden plants. Gardeners obtain the varied colors by using additions to the soil that make metal ions, like aluminum, more—or less—available to plants. In general, a more acidic soil frees metal ions. They are taken up by plants and bond to anthocyanin pigments in the saps, and the effect is a different flower color. Anthocyanins link up with other flavonoid molecules, too, creating even more color variation. Last but not least, anthocyanins are affected by pH-changes as acids or bases alter their structures. Cut red roses, when getting past their prime in a vase, often change color towards bluish tones, a result of slight changes of the pH in the cell vacuoles with ageing of the petals. It is entertaining to experiment with anthocyanin pigments: steep red cabbage in water, and then add either acid (like vinegar) or base (like baking soda) to samples of the extract and observe the different colors created!

4.6 PURPLE BETALAINS

For a long time it was assumed that anthocyanins are responsible for all the purple and purplish red colorations in plants, and for the most part this is true. But further investigations of the intense red pigments in beetroots (*Beta vulgaris*) (Figure 4.11(a)) showed that changes in pH affected these pigments differently. If you cook red beets in water and add acid or base to the colorful extracts, you can observe how the pigments from beets are much less affected by the pH changes than the anthocyanins from red cabbage.

Red beets contain *betalain* pigments that are very different from anthocyanins (see Figure 4.12).[9] They just happen to absorb similar wavelengths of light. But their molecules, although still organic and still with conjugated systems, contain nitrogen atoms. They are *alkaloids*, *i.e.* secondary metabolites with nitrogen in their structures. (We will read more about alkaloids in the next chapter on

(a) (b) (c)

Figure 4.11 Betalains in plants and fungi.
(a) Red and yellow beetroots (*Beta vulgaris*). (b) Red fruit of prickly pear cactus (*Opuntia* sp.). (c) The red cap of the fly agaric (*Amanita muscaria*) also contains betalains.

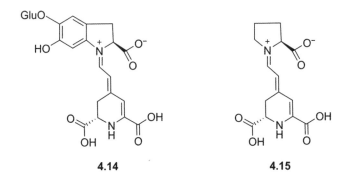

4.14   4.15

Figure 4.12 Structures of betalain pigments.
Red betanin **4.14** and yellow indicaxanthin **4.15** are both found in beetroots and in fruits of prickly pear cactus.

plant defenses.). If you compare the flavonoid structures of anthocyanins in Figure 4.9 with the examples of betalains in Figure 4.12, it is easy to detect many differences. Betanin **4.14** is the deep-red pigment in red beets. It is also found in the red fruits of prickly pear cactus (*Opuntia* spp.) (Figure 4.11(b)). There are two subfamilies of betalains: the red to red-violet betacyanins (of which betanin is an example) that absorb light in the 535–550 nm range, and the betaxanthins that are yellow-orange in color and absorb between 475–480 nm. Some specialty beets are indeed orange or yellow (Figure 4.11(a)). A comparison of the two betalain examples in Figure 4.12 shows that the conjugated system in indicaxanthin **4.15**, a yellow betaxanthin, is considerably shorter than in betanin which leads to absorption at shorter wavelengths. Betalains are water-soluble and stored in the vacuoles of plants. Just think of the deep-red beet juice that is produced when beets are cooked in water for borscht soup.

Interestingly, betalains are found only in certain plant families, all of them in the order of Caryophyllales, and there they take the place of anthocyanins. Plants either contain anthocyanin pigments or betalains—not both. This offers clues about their relationships and their evolution. Plant families that contain betalains instead of anthocyanins include the goosefoot family (Chenopodiaceae), where we find red beets, the cactus family (Cactaceae), the portulac family (Portulacaceae), and the four o'clock family (Nyctaginaceae). Just think of the hot-pink flowers of portulac roses or the garlands of purple or red bougainvillea blossoms (four o'clock family): they all contain alkaloidal betalain pigments. An interesting aside: betalains also color the red cap of the fly agaric (*Amanita muscaria*) (Figure 4.11(c)), the red and white toadstool of fairy tales.

BOX 4.1 Pigments, Scents, and Nectars Combined Attract Pollinators

Specific combinations of pigments, scents, and sugars in plants draw different pollinators.

Bird pollinators, like hummingbirds, are greatly attracted to colorful flowers, with pigments like yellow and orange carotenes or hot pink anthocyanins or betalains. Most birds have good color vision and perceive the bright colors. Hanging, tubular blossoms as in fuchsias (Figure 4.13(a)) provide good access for long bird beaks, and nectar rich in sucrose rewards the pollinators. While visiting one flower after another, birds keep pollinating the blossoms. Most birds have a poor sense of smell, and bird-pollinated flowers tend to be odorless or have faint odors only.

Bees see yellow and blue colors well and can also detect ultra-violet. Carotenoids and flavonoids like flavones and anthocyanins provide the colors. Flowers that contain UV absorbing pigments like flavones and may appear white or even-colored to humans, look more attractive to bee pollinators; they are likely to see patterns on the petals that act as nectar guides. Bees in addition are attracted by sweet smells, ample yellow pollen, and by flowers that provide a good landing area, as in the

Figure 4.13 Flower pigments and shapes for different pollinators. (a) The bright-red, tubular blossoms of fuchsia flowers (*Fuchsia magellanica* cultivar), (b) the broad-shaped, bright flower heads of mule's ears (*Wyethia mollis*) with plenty of pollen, and (c) the white, strongly scented flowers of Jimson weed (*Datura wrightii*) that open only in the evening appeal to different pollinators.

example in Figure 4.13(b)). Butterflies like bright colors, sweet odors, and a broad arrangement of blossoms that provide sugary nectar.

Flowers that are fly- or beetle-pollinated tend to have brown or maroon hues (Figures 1.33 and 3.1(b)). The colors are created by high concentrations of anthocyanins. Smells of carrion and dead fish, composed of sulfurous compounds and amines, accompany the colors and attract flies.

White and bright-yellow flowers, with flavone pigments, stand out at dusk, and attract pollinators like moths or bats. Moths frequent heavily scented, light-colored blossoms, like the Datura flower (Figure 4.13(c)). Large, light-colored flowers, with plenty of sweet nectar, invite bats.

4.7 BROWN TANNINS

Tannins are a large and diverse family of plant pigments that provide brown, grey, and sometimes reddish colors to woody tissues and to the bark of trees. The name "tannin" is derived from the German word "Tanne" for a fir. Coniferous trees like pines, firs, and redwoods have a high tannin content. So do oak trees. After rainy days, tannins contribute to the dark color of puddles underneath the trees. Old logs gradually acquire a pale, gray color, as water leaches tannins out of wood (Figure 4.14(a)). Air oxidation affects the colors, too. Tannins are also found in leaves, fruits,

Colorful Plant Pigments 109

Figure 4.14 Tannins in plants.
(a) Multicolored trunk of ancient bristlecone pine (*Pinus longaeva*).
(b) Acorns of coast live oak (*Quercus agrifolia*). (c) Oak galls on valley oak (*Quercus lobata*).

roots, and seeds. In fruits they provide a tart, astringent taste. If you ever bit into an unripe persimmon, you experienced your mouth puckering because of the high tannin content of the fruit. Native people who used acorns (Figure 4.14(b)) as a staple food knew to leach out the tannins with water first before making the seeds useful as a source of flour. Tannins are very common in plants, and we will encounter them again in the next chapter on plant defenses. Here we will take a look at their chemical structures.

Tannins are phenolics and as such are weak acids. (Compare with phenol **4.10**.). They affect the pH of soils and contribute to their acidity, especially in coniferous forests. Tannins are grouped into the *hydrolyzable tannins* and the highly complex *condensed tannins* (Figure 4.15). Hydrolyzable tannins easily dissolve in water. They are composed of relatively simple phenolic molecules, with many OH groups, and are related to gallic acid **4.16**. An example of a hydrolyzable tannin is gallotannin **4.17** found in oak galls (Figure 4.14(c)). It has a central glucose molecule that forms esters with three gallic acid groups. *Condensed tannins*, on the other hand, are composed of much larger molecules. Figure 4.15, **4.18** shows a segment of one. Note that the structure contains flavonoid ring patterns. The segment in brackets is repeated various times in different tannins. Condensed tannins provide very dark pigments in plants. They become even darker, when they bond with metal ions.

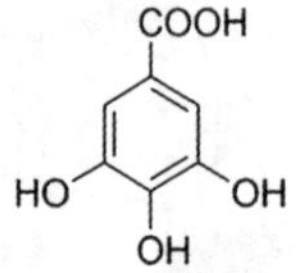

4.16

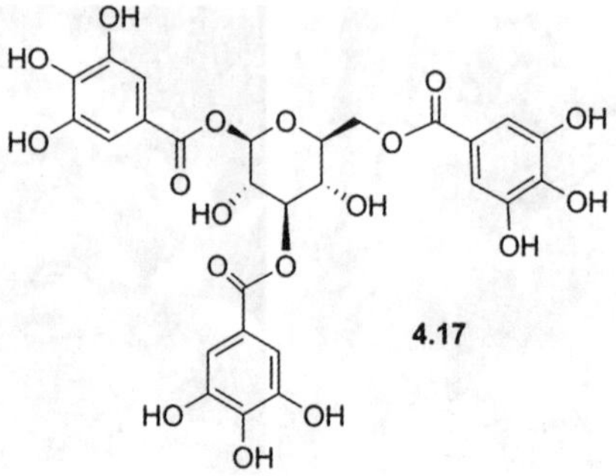

4.17

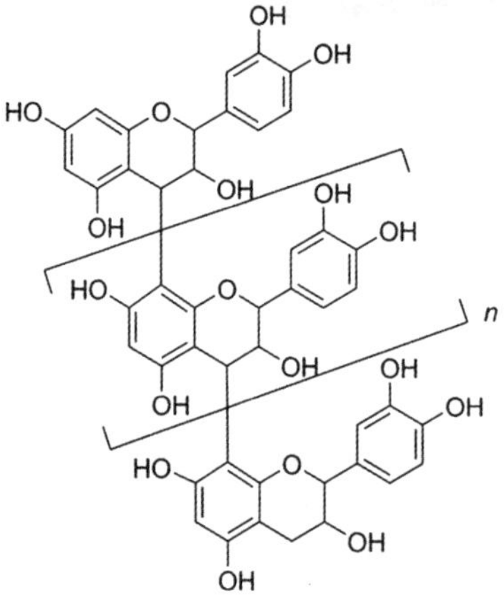

4.18

Figure 4.15 Structures of tannins.
Gallic acid **4.16** forms ester functional groups with glucose in gallotannin **4.17**, a hydrolyzable tannin from oak galls. A segment of a condensed tannin **4.18** shows flavonoid structures. (The brackets indicate the repeat pattern.).

In former times, oak galls were the source for making black ink. The tannins from the galls were extracted with water. When iron salts were added to the extract, a black ink was obtained as the metal ions strongly bonded with the tannins.

BOX 4.2 Fall Coloration

Plant leaves contain a great variety of pigment families, as illustrated with the colorful fall leaves of grape vines in Figure 1.1(b).[10] When leaves are green, chlorophyll covers up other pigments that are also present in the leaves. In fall, when production of chlorophyll ceases, and the green photosynthesis

Figure 4.16 Fall coloration in leaves.
(a) Leaves of honey locusts (*Gleditsia triacanthos*), with yellow carotenoids and xanthophylls showing, and some remaining chlorophyll. (b) Red anthocyanins in sweetgum (*Liquidambar styraciflua*). (c) Leaves of buckeye tree (*Aesculus californica*) in fall.

pigment starts to decompose, the chemically more stable yellow carotenes and flavones become visible (Figure 4.16(a)). That's when people tend to say, "The leaves turned yellow." Yet these pigments were present all along, acting as accessory pigments and also trapping harmful radiation.

Some pigments are newly formed in fall leaves. This is the case for anthocyanins, forming, for example, the bright red colors in sweetgum (*Liquidambar styraciflua*, Figure 4.16(b)). A summer with plenty of rain and also warm sunshine will enhance sugar production in the leaves, a prerequisite for good anthocyanin production. Later, if a cold snap occurs (not too cold, otherwise leaves simply shrivel up), and cool, dry fall weather sets in, the sugars act as starting materials for the biosynthesis of anthocyanin pigments. Ideal climate conditions for good fall coloration are found in the eastern United States and Canada.

The anthocyanins' role in fall coloration is still investigated. The pigments are strong antioxidants which is likely to give them a protective character. In some trees they seem to slow the formation of the abscission layer, *i.e.* the layer between tree and the stem of the leaves where leaves break off later in fall. This allows trees to continue to recover nutrients. Producing the pigments is costly for young leaves. It has been suggested that anthocyanins counteract herbivory of the tender leaves, as the red colors make insects better visible for predators.

When the weather turns very cold, or when drought conditions set in, leaves produce brown tannin pigments and dry up, as shown in Figure 4.16(c).

4.8 CONCLUSION

This chapter introduced the major families of plant pigments. We examined structures of various chlorophylls and of yellow, orange and red carotenoids, both groups being fat-soluble pigments. Among the flavonoid pigments we became familiar with the family of white and yellow flavones, then anthocyanins with many shades of blue and purple, and lastly brown tannins. Betalains are a family of alkaloidal pigments. While they also provide purple colors in plants, they are structurally unrelated to anthocyanins. Betalains are exclusively found in some plant families, all in the order Caryophyllales. In addition, there are a few smaller families of pigments, and we will encounter examples of them later.

Each pigment family introduced chemical aspects that related to the color of a pigment. The discussion of chlorophylls showed how an extension of the conjugated system through functional groups changes and deepens the absorption of light. A comparison of different carotenoids demonstrated how a shortened conjugated system leads to absorption in the UV only, an absorption that is visible to certain insects. In anthocyanins we studied the variations of colors obtained by additional functional groups. Changes in pH, and bonding with metal ions or other pigment molecules further affect the colors of anthocyanins. Similar effects are observed with tannins. The acidic, astringent properties of tannins have defensive functions and lead us to our next chapter on plant defenses.

REFERENCES

1. J. B. Harborne, *Introduction to Ecological Biochemistry*, Academic Press, London, 4th edn, 1993.
2. C. Bowsher, M. Sterr and A. Tobin, *Plant Biochemistry*, Garland Science, New York, NY, 2008.
3. J. R. Hanson, *Chemistry in the Garden*, The Royal Society of Chemistry, Cambridge, 2007.
4. P. Murphy, P. Doherty and W. Neill, *The Color of Nature*, The Exploratorium, Chronicle Books, San Francisco, CA, 1993.
5. M. Chen, M. Schliep, R. D. Willows, Z.-L. Cai, B. A. Neilan and H. Scheer, *Science*, 2010, **329**, 1318.

6. E. Breitmaier, *Terpenes: Flavors, Fragrances, Pharmaca, Pheromones*, Wiley-VCH Verlag, Weinheim, 2006.
7. S. Quideau, D. Deffieux, C. Douat-Casassus and L. Pouységu, *Angew. Chem., Int. Ed.*, 2011, **50**, 586.
8. J. Harborne, ed., *The Flavonoids: Advances in Research Since 1986*, Chapman and Hall/CRC, Boca Raton, FL, 1994.
9. T. J. Mabry, *J. Nat. Prod.*, 2001, **64**, 1596.
10. M. Archetti, T. F. Döring, S. B. Hagen, N. M. Hughes, S. R. Leather, D. W. Lee, S. Lev-Yadun and Y. Manetas, *Trends Ecol. Evol.*, 2009, **24**, 166.

CHAPTER 5

Poisons and Other Defenses

5.1 INTRODUCTION

Plants have had to defend themselves since their emergence millions of years ago. Myriads of insects assail them, snails feed on them, and larger herbivores devour fresh, green leaves and juicy stems. Plants also need to fight against attacks by microorganisms and fungi. Competing plants may cut off light and water supply. In addition, plants have to be able to survive in their specific habitats, in deserts, in alpine areas, or in environments with large fluctuations in temperature and rainfall. Being mostly anchored in place, plants have had to adapt to cope with the challenges. It is truly astonishing that plants are so abundant in spite of all the threats to their survival.

Elaborate defense systems make it possible for plants to live on, and many of the defenses entail amazing plant chemistry. When we think about plant defenses, a thorny rose may come to mind, or a cactus full of spines, or the tough leaves of agaves rimmed with spines (Figure 5.1(a)). Indeed, most plants have structural defenses: thick skins, thorns, and sharp spines protect tender shoots from browsing animals. Tough skins of leaves help reduce water loss. Fine, white hairs reflect intense sunlight. Here we focus on the chemical defenses that plants have evolved. Chapter 2 mentioned some chemical adaptations that provide better tolerance of freezing

---

The Chemistry of Plants: Perfumes, Pigments, and Poisons
Margareta Séquin
© Margareta Séquin 2012
Published by the Royal Society of Chemistry, www.rsc.org

Figure 5.1 Structural and chemical defenses in plants.
(a) Leaves of desert agave (*Agave shawii*) have sharp spines and tough-skinned leaves. (b) Foliage of European yew (*Taxus baccata*) is highly toxic to horses, but berries can be ingested by birds without harm. (Photo by Ruth Marent.).

temperatures, such as high sugar content of plant saps or unsaturation in fatty acid components of membranes. This chapter addresses the chemical defenses that plants have evolved to ward off sucking and browsing animals, to fight against fungal attacks, and to keep other plants from growing too close. The chemical defenses can be relatively gentle, like distinctive strong odors of leaves that keep animals from browsing on them, or they may be in the form of sticky resins that fight insect attacks. Bitter or irritating saps can discourage larger animals from eating leaves and branches. In addition, plants have evolved potent toxins that seriously injure or kill organisms that try to feed on plant parts.

Plant defensive substances do not affect all organisms the same way.[1] We need to keep in mind that the concept of "toxin" or "poison" is relative: the quantity (or *dose*) ingested determines how harmful or toxic a compound is, but differences in animal metabolism can also be a factor. A plant compound that is detrimental to some animals may be harmless to others. A substance that kills small insects may not affect larger animals. There are differences in susceptibility even among vertebrates. In addition, the various parts of plants often contain different concentrations of defensive compounds. For example, all parts of the European yew (*Taxus baccata*) (Figure 5.1(b)) contain a mixture of toxic alkaloids, called taxine, except for the fleshy part of its red berries. Birds can eat the berries without harm (and spread the seeds) while horses that feed on clippings from hedges of yew become seriously ill or die.

The number of defensive compounds in plants is huge, and new ones keep being discovered. Their chemical structures are highly diverse and include unrelated classes of compounds. Their degree of harmfulness to animals or humans cannot easily be determined by merely studying the chemical structures. Comparison with known defensive compounds and their activities sometimes points to potentially toxic properties—not necessarily with correct results. The feeding behavior of animals with respect to plants is often a major lead to investigations on chemical plant defenses.

This chapter provides an overview of major classes of defensive plant compounds and introduces some representative examples along with plants that contain them. While studying the different types of chemical defenses, we will apply our knowledge of organic structures. We will encounter some earlier compound families, like terpenes and phenolics, but here in their roles as plant defenses. The sequence of chapter sections follows increasingly more potent effects of plant compounds on other organisms. The chapter starts out with fairly harmless defenses, such as odorous volatile compounds in leaves. Plant defenses in the form of viscous, distasteful, or irritating chemicals follow. Some plants produce toxic cyanide when their leaves or seeds are bitten into and are described next. Soapy saponins and cardiac glycosides that affect the activity of the heart muscle introduce the structures of steroids. The chapter concludes with the plant bases called alkaloids, a family of defensive plant compounds that contains many well-known toxins. A defense in the form of sharp crystals in leaves is described in Box 5.1. Some insects have learned not only to cope with chemical plant defenses, but use them for their own defense. Some examples are shown in Box 5.2.

Looking further ahead to our later chapter on plant medicines, let us recall that plants, in their fight against unwanted organisms, have supplied us and keep providing us with a large number of compounds that are useful in our own battle against human diseases.

5.2 STRONG SCENTS IN LEAVES

A first line of chemical defense in plants may be a strong odor in leaves or needles, or in the peel of fruits. The smells are provided by

Figure 5.2 Plants with strong odors in their leaves.
(a) Lemon balm (*Melissa officinalis*) in the mint family. (b) Branches of blue-gum eucalyptus (*Eucalyptus globulus*).

essential oils. In fact, some of the same compounds that we encountered as fragrant attractants in flowers can act as repellents in leaves. Limonene **3.4**, the smell of lemons, contributes to attractive flower scents. But higher concentrations in citrus leaves and in the peel of lemons act as repellents towards insects and larger animals. We are quite familiar with strong smells in leaves when we think of herbs like sage, thyme, or various mints (Figure 5.2(a)). Animals in general do not favor a diet of strongly scented leaves. The smells of many herbs, like rosemary or bay leaves, are potent insect repellents.[2] For that reason, they are sometimes still added to bags of flour. (In earlier times the strong scents also covered up the taste of spoiled food!)

Essential oils with strong odors are common in plants that grow in deserts or in habitats with seasonal droughts. The numerous species of eucalyptus (Figure 5.2(b)) are native to arid areas of Australia. Note that many culinary herbs originate from areas with a Mediterranean climate where plants have to be able to cope with many months without rain. Without the distinct odors, young shoots would be vulnerable to animals keen on feeding on the juicy leaves.

Monoterpenes, *i.e.* isoprenoid compounds with ten carbon atoms, are dominant in the mixtures of essential oils that provide odors to leaves (Figure 5.3). Many names of the compounds reflect their plant origin: menthol **5.1** is found in mint plants, eucalyptol (also known as 1,8-cineole) **5.2** in eucalyptus leaves, and α-pinene **5.3** in pine needles.[3,4] Other plants contain these terpenes as well: eucalyptol is also found in the essential oils of rosemary (*Rosmarinus officinalis*), and eucalyptus oil contains α-pinene among its many other components. Odorous terpenes contribute to mixtures of volatile compounds that together compose the smell of the leaves of a plant. Different types of mints, like lemon mint or spearmint, have different compositions of their essential oils; we can easily distinguish the leaves by their smells.

Many defensive plant compounds, including defensive odors, have chiral centers and are composed of asymmetric molecules. Their asymmetries are pointed out with the usual wedge and dash representations in the figures. (Not all the chiral centers are pointed out, though, for simpler presentations.) Sometimes only one of several possible stereoisomers is found in nature. For example, menthol **5.1** is naturally found only as the isomer shown in Figure 5.3. α-Pinene **5.3** exists as two mirror images or enantiomers; they are shown in Figure 5.3. Interestingly, the volatile oils of North American pines contain one of the enantiomers of α-pinene, whereas most pine oils of European origin have the other enantiomer. Equal amounts of two enantiomers are called *racemic mixtures*, and oils from *Eucalyptus* spp. contain racemic mixtures of the two mirror images of α-pinene.

Oily, water-repelling terpenes not only provide strong scents in warm air, but also form protective layers on leaves that help cut down on water loss for plants. Their hydrocarbon nature makes them highly flammable. Plants rich in terpenes, like pines or eucalyptus, burn very well in a forest fire! Read later in this chapter about the role of terpenes as *allelopaths*, *i.e.* as compounds that inhibit the growth of other plants.

Phenolic compounds are another class of defensive plant smells. They often exhibit pungent odors. The compound phenol **4.10** itself

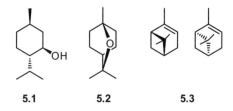

5.1   5.2   5.3

Figure 5.3   Terpenes as defensive plant odors.
Menthol **5.1**, eucalyptol (1,8-cineole) **5.2**, and α-pinene **5.3** are common monoterpenes in plant odors. The two enantiomers of α-pinene are shown.

is a crystalline compound with a sharp smell that reminds us of disinfectants. (It is not found in plants.) Similarly many phenolic compounds contribute sharp plant odors, especially in resinous layers on leaves, as will be shown in the next section.

5.3 STICKY RESINS

Another line of plant defense can be in the form of viscous resins that discourage attacks by insects or of gums that seal plant wounds.[5] The terms "resin" and "gum" are often used interchangeably. Both describe elastic, sticky saps or *exudates* from plants. Strictly speaking, plant resins are non-polar, and their structures are largely hydrocarbons. (Resins burn well, too.) Gums include more polar substances and consist of large carbohydrate structures. True gums will dissolve in water. They are usually formed as a response to injury, like the sticky saps that form when a cherry or plum tree is being trimmed. Common plant names like "gum tree", or "sweet gum", or "tarweed" allude to plants with sticky exudates—which, technically speaking, are all water-repelling resins. Later we will encounter other defensive plant saps, in the form of milky latex. Here we focus on hydrophobic plant resins.

Conifers, like pines and firs, are well-known for their resin production, especially in their bark and in cones (Figure 5.4(a)). When the bark of a pine tree is injured, water-repelling resins ooze

(a)   (b)   (c)

Figure 5.4   Plant resins.
(a) Cone of bristle cone pine (*Pinus longaeva*) exuding resin. (b) Creosote bush (*Larrea tridentata*), a common bush in deserts of North America's southwest, has resinous branches and leaves. In rainy weather the resins give off a characteristic, sharp smell of the phenolic NDGA (nordihydroguaiaretic acid). (c) Amber, a fossilized resin.

out and protect the wound. Many resins are complex mixtures of terpenes. Fragrant monoterpenes provide the typical smells of resins, as in pines. Larger terpene molecules, with fifteen or twenty carbon atoms, are also part of the mixture. The larger molecules do not evaporate like the monoterpenes, but become a viscous, almost liquid mass at higher temperatures. The resins do not dissolve in water, as we can experience when we try to wash off the sticky mass after touching the bark of a pine tree.

Pine resins can be commercially treated with steam to produce turpentine oil. The hot steam carries off the more volatile, smaller molecules from the resins, producing the liquid mixture with a distinct smell. The remaining sticky residue or *rosin* consists mainly of the less volatile *diterpenes*, *i.e.* isoprenoids with twenty carbon atoms. A major component of rosin is the diterpene abietic acid **5.4** (Figure 5.5). Four isoprene units compose abietic acid. Can you find them? Rosin (also called colophony) has many different applications. For instance, musicians that play string instruments use it to treat their bows, to make the bow hair grip the strings.

Other resins are composed of complex phenolic compounds. They can be mixed in with terpene resins. Creosote bush (*Larrea tridentata*) (Figure 5.4(b)) is a common, highly drought tolerant desert shrub in deserts of southwestern North America; related *Larreas* are found in Central and South American deserts. Their leaves are covered by phenolic resins that help reduce water-loss from leaf surfaces. Creosote branches emanate a distinct, sharp odor that is especially noticeable after a rain. The pungent odor is

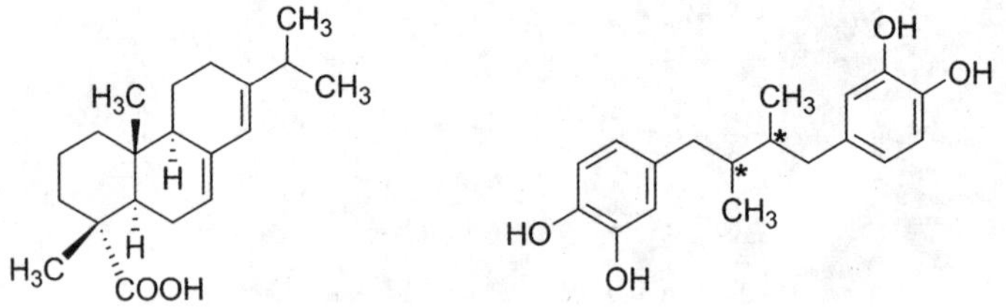

Figure 5.5 Terpenes and phenolics in resins.
Abietic acid **5.4**, $C_{20}H_{30}O_2$, is a common diterpene found in plant resins. NDGA (nordihydroguaiaretic acid) **5.5** is a phenolic compound that contributes to the pungent smell of resins on the leaves of creosote bush (*Larrea* spp.).

produced by nordihydroguaiaretic acid (NDGA) **5.5** that is part of the resinous, hydrophobic mixture. Can you see that NDGA is a phenolic compound? Like all phenolics, NDGA is a potent antioxidant, which gives it an additional protective role in plants. Incidentally, its action as an antioxidant makes it a useful additive to inhibit oxidation of fats and oils in cosmetics.[6]

Under the right conditions, plant resins can fossilize into amber (Figure 5.4(c)).[7,8] Special conditions must be present to lead to a polymerization of the resins. The formation of a fossilized form is a unique property among secondary plant metabolites. Most ambers are derived from terpenoid resins from trees that were carried down by streams; the resins, being mostly hydrocarbons, floated on water. As sediments buried the resins, polymerization and hardening began, and amber was formed. The chemical composition of the resins can give us information about the origin and the age of a piece of amber. Moreover, enclosures of insects and seeds provide insight into the fauna and flora of ancient time periods. Radiocarbon dating determined ambers to be more than 40 000 years of age.

5.4 DEFENSIVE SULFUROUS SMELLS AND TASTES

If snails, insects and other browsers are not deterred by strong smells or sticky surfaces and proceed with feeding on plants, then a next line of defense may be a repelling taste of leaves, roots, or fruits. Plants in the cabbage family (Brassicaceae), like broccoli, cabbage, or mustard (*Brassica* spp., Figure 5.6(a)), store defensive mustard oils called *glucosinolates* **5.6** in some of their cells (Figure 5.7). When animals start chewing on the plants, cell walls are broken, and enzymes from neighboring cells start reacting with the mustard oils. The enzyme myrosinase breaks the glucosinolates into smaller, volatile molecules that have an acrid smell and sharp taste. You can experience this when crushing the flower buds of a mustard plant: a distinct smell is produced, familiar to us from mustard used in foods. When the glucosinolate sinigrin **5.7**, a common mustard oil in cabbages, is disintegrating, a sharp-smelling, volatile compound with the name of allyl isothiocyanate **5.8** is formed (Figure 5.7). (The syllable "thio" means that sulfur is part of the molecule.) Such strong smells, accompanied by sharp tastes, are strongly repelling to small insects and slugs.

Figure 5.6 Plants that produce defensive sulfurous smells.
(a) Mustard plant (*Brassica juncea*). (b) Garlic, onions and leeks (*Allium* spp.).

Many plants used for human consumption are in the mustard family. This is probably not a coincidence as the glucosinolates provide the growing plants with natural chemical defenses towards pests. When cabbages or broccoli are cooked, some of the glucosinolates are destroyed or transformed into other sulfurous volatiles. Humans, in general, do not seem to be deterred by the sulfurous odors (although some adults and many children may think otherwise!). Some specialized insects are not repelled by the smells and tastes either. Through evolution they have learned to cope with the deterrents: they either sequester the compounds or rapidly metabolize them. An example in case is the cabbage white butterfly (*Pieris* spp.) and its caterpillar, a major pest in agriculture. The caterpillars are seemingly undeterred by the glucosinolates in cabbage and even take advantage of the nutritional value of the glucose formed as a metabolic product of the enzymatic interaction with glucosinolates.[9]

A similar enzyme action is required to produce the distinct odor and taste of onion and garlic plants (*Allium* spp., Figure 5.6(b)).[10] Unharmed bulbs of garlic or onions have no smell. The odor develops only when the bulbs are cut. In the case of onions, this is also the moment when our eyes start tearing up. When bulbs or leaves of plants of the genus *Allium* are cut or bitten into, the enzyme alliinase begins to react with alliin **5.9**, an odorless compound. The reaction breaks alliin into smaller volatile molecules that have strong smells and tastes. Allicin **5.10** provides the typical odor of garlic. Another volatile break-down product from alliin is

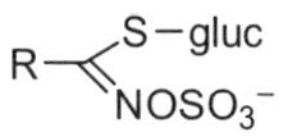

Poisons and Other Defenses 123

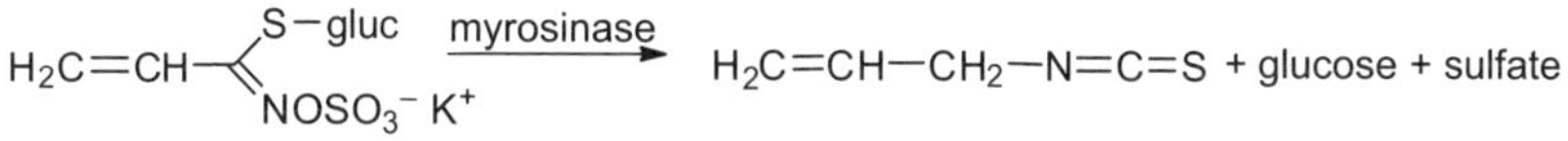

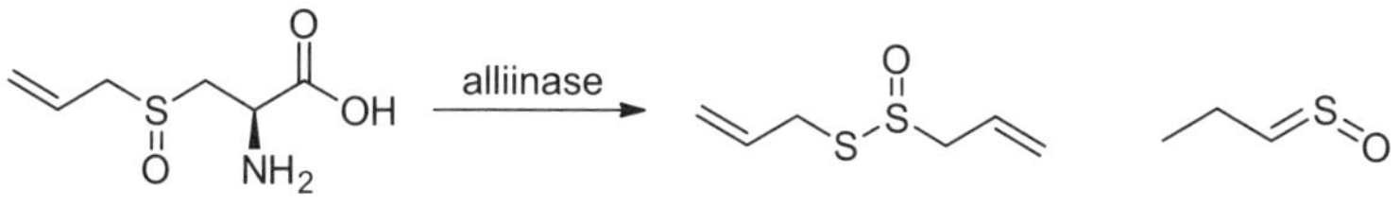

Figure 5.7 Glucosinolates and other defensive sulfurous plant compounds. The common structure of glucosinolates **5.6** is found in sinigrin **5.7**, a defensive compound in plants of the cabbage family (gluc: glucose group). Enzymes from neighboring cells break the glucosinolate into allyl isothiocyanate **5.8**, a volatile compound with a sharp smell and taste. Onions and garlic contain alliin **5.9** which, upon enzyme action, is converted into allicin **5.10**, the typical smell of garlic, or into the tear-producing thiopropanal-S-oxide **5.11** from onions.

responsible for the tear-producing (or *lachrymator*) action of cut onions; its name is thiopropanal-S-oxide **5.11** (Figure 5.7). Any plant in the genus *Allium* (like the wild onion shown in Figure 1.3(b)) will produce a typical onion smell when its leaves are crushed because of the volatile sulfurous compounds that are formed.

5.5 SOUR ACIDS

Other repelling tastes may discourage animals from feeding on plants, namely sour or bitter tastes. We need to keep in mind that the assignment of sour or bitter necessarily relates to the human experience.[11] But observations of herbivores and their feeding preferences provide much information about repellents in plants. Range animals avoid certain plants unless under stress. Insect pests have preferences towards some plants while hardly attacking others. Again, compounds that are distasteful to some animals may have no deterring effects on others.

Organic acids are mostly responsible for what we consider a sour taste in plants. Their sourness keeps us from eating unripe fruits. Other vertebrates react the same way. Sour organic acids help fruits and their seeds reach maturity before seeds are distributed. During the ripening process, the acids are transformed into volatile esters that produce attractive odors or are degraded while the sugar content of the fruits increases. (Lemons are an exception!) Acids in stems and leaves, like in rhubarbs, discourage browsing animals and often have harmful effects on the browsers, too.

Organic acids are widely distributed in plants. Many of their names remind us of the plants they are found in, like oxalic acid **5.12** (in sour grass or *Oxalis* spp.), malic acid **5.14** (in apple or *Malus* spp.), citric acid **5.15** (in citrus plants), or salicylic acid **5.16** (in willow or *Salix* spp.) (Figure 5.8). But other plants contain these acids as well. Remember that the typical functional group of organic acids is the carboxylic acid group, COOH. They are relatively weak acids, with a pH around 4 or 5 when dissolved in water. Spinach gets its slightly acidic taste from oxalic acid. Its sodium salt is found in sour grass or sorrel (*Oxalis*), and we can detect its sour taste when we nibble on some of its leaves. Rhubarbs, especially their leaves, contain a high concentration of

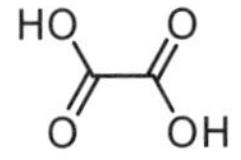

$$HOOCCOOH_{(aq)} + Ca^{2+}_{(aq)} \longrightarrow Ca(COO)_2 + 2H^+_{(aq)}$$

HOOCCOOH
5.12

5.13

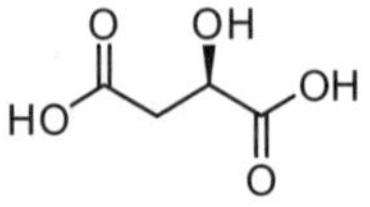

5.14 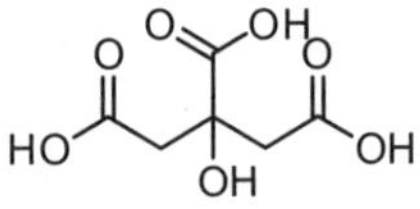   **5.15** 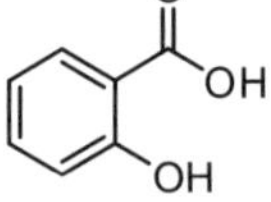   **5.16**

Figure 5.8 Organic acids in plants.
Oxalic acid **5.12**, shown as a line structure and as a condensed structural formula, forms water-insoluble calcium oxalate **5.13** with calcium ions. Malic acid **5.14**, citric acid **5.15**, and salicylic acid **5.16** are other common organic acids in plants.

oxalic acid. The potentially harmful effect of oxalic acid relates to the fact that the compound forms strong bonds with essential ions like calcium and iron ions, forming water-insoluble oxalates. In spinach, iron ions are tied up as insoluble iron oxalate, making most of the iron in the vegetable unavailable as a nutrient.[12] Read in Box 5.1 about plants that have sharp crystals of calcium oxalate **5.13** in their leaves.

BOX 5.1 Sharp Oxalate Crystals

Oxalic acid easily bonds with calcium ions (Figure 5.8), forming water-insoluble calcium oxalate **5.13**. (A very painful manifestation of calcium oxalate in the human body is in the form of kidney stones.) In some plants, calcium oxalate forms dagger like crystals called *raphides*, from the Greek word for needle.[13] Calcium oxalate crystals are typically found in the leaves of plants in the Arum family. Examples are Calla lilies (*Zantedeschia aethiopica*, Figure 5.9(a)) or many tropical plants that we keep as house plants, *e.g.* all the philodendrons, or the attractive houseplant dumb cane (*Dieffenbachia* spp., Figure 5.9(b)). Care needs to be taken to keep pets and small children from chewing on the plants. The needlelike crystals of calcium oxalate would injure the browser and cause severe inflammations and swelling of the tongue. *Dieffenbachia* got its common name of dumb cane for a reason! Taro (Figure 5.9(c)), a tropical food

(a) (b) (c)

Figure 5.9 Plants with raphides in their leaves.
(a) Calla lily (*Zantedeschia aethiopica*). (b) Dumb cane (*Dieffenbachia* sp.). (c) Taro plant (*Colocasia esculenta*).

plant and source of starch, is also a plant in the Arum family. Only lengthy cooking and steeping of taro root will destroy the sharp calcium oxalate crystals and make the roots edible.

5.6 BITTER TASTES

Compounds that taste bitter to us have greatly diverse structures, as demonstrated with the examples of bitter-tasting plant compounds in Figure 5.11. Their structures are complex and may look bewildering. Yet, we can find relations to compound families that we have encountered earlier. All parts of plants can contain bitter compounds, but they are usually concentrated in one section of a particular plant. The roots of yellow gentian (*Gentiana lutea*, Figure 5.10(a)), the bark of the *Cinchona* tree that provides quinine, the unripe fruits of persimmon trees, or the leaves of the coca plant (*Erythroxylum coca*) all contain bitter substances. Many bitter plant compounds have found human applications.

First let us focus on *tannins*, a family of pigments that were introduced in the previous chapter (Figure 4.15). Here we encounter them in their role as plant defenses. Tannins are a large family of bitter-tasting and astringent compounds. Plants or plant parts that are high in tannin content are generally avoided by animals. Plant names like bitterroot (*Lewisia rediviva*) or alum root

Figure 5.10 Plants with bitter compounds.
(a) Yellow gentian (*Gentiana lutea*). (b) Bark of coast redwood tree (*Sequoia sempervirens*). (c) Inner part of grapefruit peel.

Poisons and Other Defenses 127

(*Heuchera* spp.) allude to their high tannin content. In small amounts, on the other hand, tannins add distinct flavor to many of our foods, like wines.

Tannins are phenolics and as such are acidic. They strongly bond to proteins through hydrogen bonding and more complex molecular interactions. This causes peptides and proteins to *coagulate*, forming clots that precipitate out. This means that tannins make proteins less available to animals. A high tannin content of cattle feed reduces the digestibility of proteins for the animals, a highly undesirable effect. Tannins are thus known as *antinutrients*. Tannins are harmful to insects. Not only do they bond to enzymes, but they also form toxic compounds in their guts. The effect of tannins on proteins is used in the process of tanning leather. The gelatinous fresh animal hides are treated with mixtures that are high in tannin content. The proteins in the hides are precipitated out and made stable as a result of the tannins acting on them. Tannins are antifungal, too. Therefore, wood from trees that are high in tannin content is highly desirable as lumber as it resists fungal rot. Tannins give trunks of coast redwoods (*Sequoia sempervirens*) (Figure 5.10(b)) their red-brown color.

Plant compounds that exhibit bitter tastes (again from the human perspective) can have very different chemical structures. Think of the bitter taste of the white, soft inner part (called the albedo) of grapefruit peel. The taste is imparted by naringin **5.17** (Figure 5.11), a flavonoid glycoside with a very bitter taste. (The carbohydrate group is noted as "glyc" in the structure of naringin.)

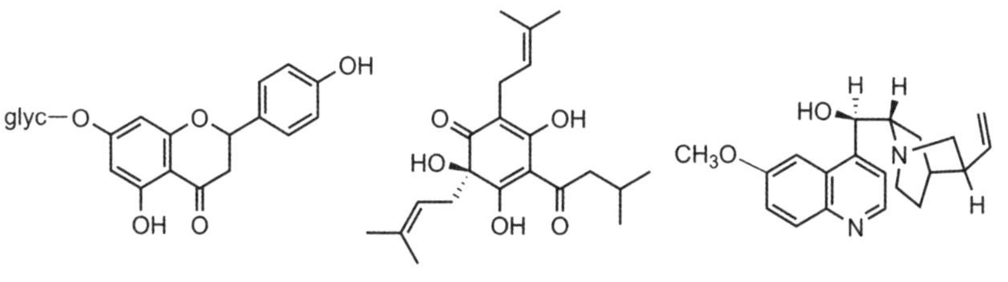

Figure 5.11 Structures of bitter-tasting plant compounds.
The great diversity of bitter tasting compounds is represented by naringin **5.17**, a flavonoid glycoside, humulone **5.18** derived from terpenoid structures, and quinine **5.19**, an alkaloid. (glyc: carbohydrate group).

Interestingly, the molecule without the carbohydrate group attached, *i.e.* its *aglycon*, is not bitter. This is made use of commercially: small amounts of the enzyme naringinase are added to packaged grapefruit juice to lower its bitterness.[14] The resin of hop plants (*Humulus lupulus*) contains several bitter compounds, among them the compound humulone **5.18**. It is important for the production of taste in beer. Notice the isoprene units in humulone.

For a very different structure of a bitter-tasting compound, look at the structure of quinine **5.19**. It is an example of an *alkaloid*, *i.e.* an organic compound that is a secondary plant metabolite with nitrogen in its structure. Alkaloids often have complex ring structures (which is certainly the case for quinine). They tend to have a bitter taste and often act as defensive substances in plants. Alkaloids frequently have potent physiological effects on animals and humans. The last section in this chapter will elaborate more on alkaloids, and we will encounter them again in the chapter on human uses. Quinine is an intensely bitter compound that can be extracted from the bark of the tropical trees of *Cinchona* spp. The compound has many medicinal applications; probably best-known is its use as an antimalarial. Its bitter taste is quite familiar from tonic water or gin and tonic.

5.7 MILKY SAPS

Milky saps or *latex* are another form of plant defense. Many of these saps act as potent insect repellents. Plant names like milkweed (*Asclepias* spp.) or wolf's milk (*Euphorbia* spp.) allude to stems and leaves that contain latex. Some very common plants contain milky saps. Next time you are weeding out dandelions (*Taraxum* spp., Figure 5.12(a)), notice the latex that shows up wherever the plant is broken off, especially at the base of the stems. With increasing age of the plants, dandelions that are tasty as salad greens in early stages develop bitter ingredients as part of their latex. Similarly, leaves of lettuce (*Lactuca* spp.) contain latex, especially types of the "romaine" variety. The Latin name *Lactuca* is in fact derived from the word for milk. Lettuce that is allowed to grow too long also develops a bitter taste. Bitter or irritating substances are often part of milky saps.

Latex is a mixture of water with tiny rubber particles. Other water-insoluble solids and liquids are finely dispersed and suspended

Poisons and Other Defenses 129

Figure 5.12 Plants with milky saps.
(a) Dandelions, when stems are broken, produce milky latex. Plants of the Euphorbia family, like (b) the decorative houseplant croton (*Codiaeum variegatum*), and (c) poinsettias (*Euphorbia pulcherrima*), contain latex, often with irritants. (5.12(b): photo by Joan Hamilton.)

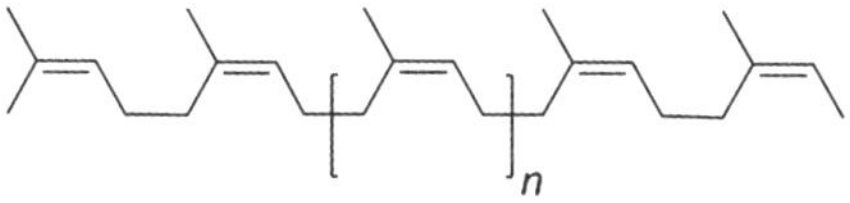

5.20

Figure 5.13 Structure of natural rubber.
In natural rubber **5.20** the isoprene units are arranged *all-cis*.

in the mixture, creating a milky appearance. Water-insoluble oils in emulsion with water contribute to latex, too. Natural rubber or caoutchouc **5.20** is a polymeric terpene (Figure 5.13).[15] Its structure consists of many repeating isoprene units. Note that the units are all connected in a specific orientation with respect to the double bonds: natural rubber has an *all-cis* arrangement of its monomers. Other rubber-like materials have different arrangements. This gives them other properties, like different elasticity. Latex from the rubber tree (*Hevea brasiliensis*) is the major plant source for the commercial production of rubber. (Nowadays most rubber is made synthetically.) Other plants with rubbery latex have been explored as alternative sources, too, like the rubber rabbitbrush (*Ericameria nauseosa*) that grows in arid regions of western North America, and even some types of dandelions.[16]

Plants of the Euphorbia family, also known as spurges or wolf's milk, are notorious for their irritating milky saps.[17] The colorful houseplant croton (Figure 5.12(b)), also in the Euphorbia family,

Figure 5.14 Plants with irritants.
(a) Stinging nettle (*Urtica dioica*). (b) Poison oak (*Toxicodendron diversilobum*) in fall. (c) Mala mujer (*Cnidoscolus angustidens*) in the Euphorbia family. (Photo by Joan Hamilton.)

has badly irritating latex and needs to be kept away from pets and small children. Mala mujer (*Cnidoscolus angustidens*, Figure 5.14(c)), a tree-like plant from Mexico and southwestern North America, is another plant with irritating saps in the Euphorbia family. Its Spanish common name means "bad woman", and the Latin name *Cnidoscolus* comes from Greek for "nettle" and "prickle". Read more about plant irritants in the next section. Poinsettias (*Euphorbia pulcherrima*, Figure 5.12(c)) that often decorate houses at holiday time produce latex when their leaves are broken, although their saps are not particularly irritating. Milkweeds (*Asclepias* spp.) are known for their latex that contains cardiac glycosides. These plant defenses will be addressed later in this chapter.

Let us end on a more harmless note relating to latex: with chewing gum. The source of the gum part was originally chicle, a sweetish rubbery material obtained from the coagulated latex of the sapodilla tree (*Manilkara* spp.) from Mexico and Central America.

5.8 IRRITANTS

Plant substances that cause painful inflammations and allergies in humans and animals can have greatly diverse chemical structures.[18] Probably one of the best-known plants for its irritating action is stinging nettle (*Urtica dioica*, Figure 5.14(a)). Leaves and stems of the plants have fine, silica-tipped hairs called *trichomes* that act like

Poisons and Other Defenses 131

hypodermic needles. They sting the browser and inject a mixture of neurotransmitters containing acetyl choline **5.21**, serotonin **5.22**, and histamine **5.23** (a name known from antihistamines that counteract allergenic reactions). Mixed in is also formic acid **5.24**, a simple carboxylic acid that has a sharp smell and is strongly irritating (Figure 5.15). Ants use it as a defense mechanism, too.

Poison oak (*Toxicodendron diversilobum*, Figure 5.14(b)) and poison ivy (*Toxicodendron radicans*) are North American plants that are infamously known for their irritating oils. They seem to affect humans only. All parts of the plants contain oily phenolic compounds, called urushiols. One of the urushiols **5.25** shows their phenolic structure and their typical long hydrocarbon sidechain (different urushiols have slight differences there). The nonpolar properties of urushiols make them well soluble in the oily layer on skin and are rapidly absorbed.

As mentioned earlier, the latex of Euphorbias, with plants like croton (Figure 5.12(b)) or mala mujer (Figure 5.14(c)), often

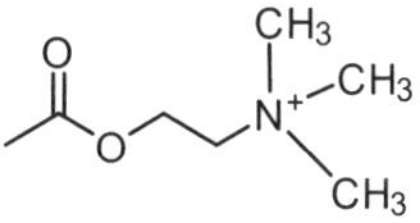

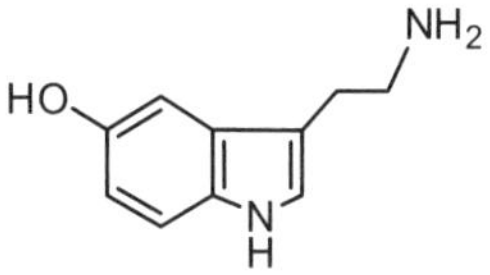

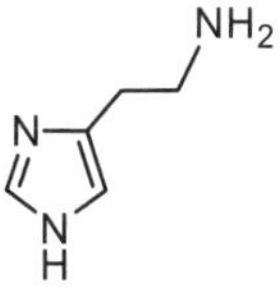

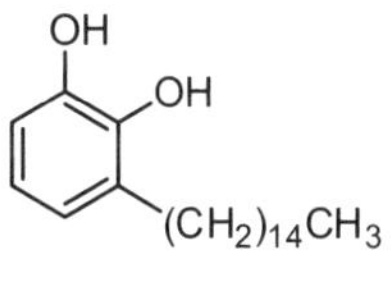

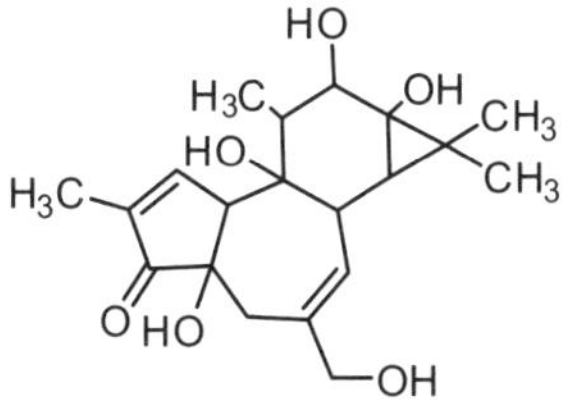

Figure 5.15 Diverse structures of plant irritants.
The neurotransmitters acetyl choline **5.21**, serotonin **5.22**, and histamine **5.23**, are found in stinging nettles (*Urtica dioica*). Formic acid **5.24** is also part of the irritating mixture. Urushiol **5.25** is a non-polar, phenolic compound from poison oak (*Toxicodendron diversilobum*). Phorbol **5.26**, a diterpene, is a carcinogenic irritant found in the latex of *Euphorbia* spp.

contains compounds that are irritating. Phorbol **5.26** and its esters are common irritants in these plants. Many studies have been focused on phorbol and its esters because they are strongly carcinogenic.[19] Note that phorbol has a complex structure with several rings. The molecule has numerous chiral centers. Can you find some of them? The molecular formula of phorbol is $C_{20}H_{28}O_6$. It is a terpene (specifically a diterpene). With careful examination of the structure you can make out some of the isoprene units.

5.9 GROWTH-REPRESSING ALLELOPATHS

Plants need to defend themselves for survival, and not only from animals. Neighboring plants crowd in and compete for water, nutrients, and sunlight. Many plants contain defensive chemical compounds that repress the growth of nearby plants, even of the same species, and keep seeds from sprouting. The chemicals can be exuded from the roots or are released from fallen leaves. The inhibition of plant growth by chemical compounds produced by another plant is known as *allelopathy*.[20]

Black walnut trees (*Juglans nigra*) have long been known for their powerful allelopaths that cause other plants around them to have stunted growth or to wilt and die. The effective inhibitant is juglone **5.27** (Figure 5.16), a phenolic compound found as its glycoside in the plants. When leaves fall to the ground and interact with moisture from the soil, the glycosides are hydrolyzed and form the active allelopath. Juglone inhibits the growth of many types of plants that grow nearby. Therefore it is sometimes used as a natural herbicide. As an aside, juglone is an intensely yellow compound that stains hands when the green husks surrounding walnuts are peeled off. (Note the conjugated double bonds.)

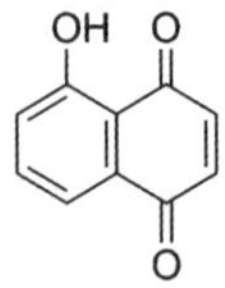

5.27

Figure 5.16 An allelopathic compound.
Juglone **5.27** is a yellow compound with a phenolic structure.

Poisons and Other Defenses

Plants that grow in arid habitats or in seasonally dry areas are in tough competition with other plants, and allelopathy is one of the mechanisms to keep other plants from growing nearby. We encountered creosote bush earlier (Figure 5.4(b)), a plant that is highly adapted to desert conditions. Its phenolic, odorous compound NDGA **5.5** is suspected to be a potent allelopath.

Plants that are high in volatile terpene content can inhibit the growth of neighboring plants, too. Terpenes from leaves and root exudates of eucalyptus trees have been found to effectively keep other plants from sprouting. Plants in shrub communities found in Mediterranean climates, like the chaparral of California, are sometimes surrounded by patches of bare soil. A well-studied example is purple sage (*Salvia leucophylla*), a shrubby sage plant of the chaparral.[21] The suspected mechanisms involve its terpenes that effectively keep other plants from crowding in.

5.10 HARMFUL CYANIDES

If we have the bad luck to bite into a bitter almond, we experience its bitterness and may even get a hint of almond smell. Biting into an apple pit (Figure 5.17(a)) or into the soft inner part of an apricot or peach pit leads to a similar, although milder experience. The soft seeds—the future of the plants—are protected by hard casings, but also chemically by *cyanogenic glycosides*. The name of these plant toxins sounds ominous enough: these glycosides generate hydrogen cyanide (HCN) when cell walls between neighboring cells are broken down, *e.g.* through crushing or grinding of plant parts.[22]

(a) (b) (c)

Figure 5.17 Cyanogenic glycosides in plants.
Cyanogenic glycosides are found in (a) pits of apples, (b) leaves of Catalina cherry (*Prunus lyonii*), and (c) leaves of white clover (*Trifolium repens*).

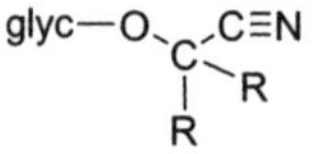

$$\text{glyc}-O\underset{R}{\overset{C\equiv N}{\underset{|}{C}}}R$$

5.28

$$\text{gluc}-O-\text{gluc}-O\underset{Ph}{\overset{C\equiv N}{\underset{|}{C}}}H \xrightarrow{\text{enzyme}} HCN + Ph\text{-CHO} + \text{glucose}$$

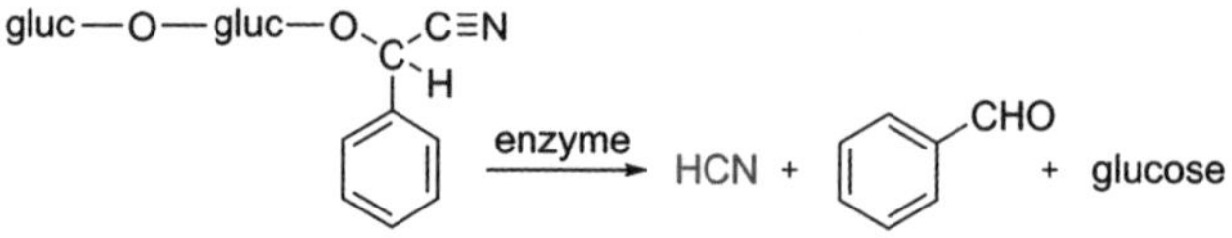

5.29 **5.30**

Figure 5.18 Structures of cyanogenic glycosides. The general structure of cyanogenic glycosides **5.28** (glyc: carbohydrate group) is seen in the example of amygdalin **5.29** (gluc: glucose). Enzyme action leads to the formation of hydrogen cyanide (HCN) and benzaldehyde **5.30**, as well as glucose.

Specific enzymes react with the glycosides, and hydrogen cyanide gas is formed as a product. Cyanogenic glycosides **5.28** have relatively simple organic structures (Figure 5.18). A representative example is amygdalin **5.29**. It can be found in seeds, leaves, and tree barks of plants in the rose family (*Rosaceae*), especially in the genus *Prunus*. Peach trees, cherry trees, and almond trees (*Prunus dulcis*) all belong to this genus. When amygdalin reacts with the enzymes, hydrogen cyanide and benzaldehyde **5.30** are formed. Both compounds have a smell of bitter almonds. If leaves from a Catalina cherry tree (*Prunus lyonii*) (Figure 5.17(b)) are crushed, this smell is quite noticeable. The low dose of cyanide that is released does not affect humans. But snails and insects that try to feed on leaves are repelled or harmed by the release of cyanide.

Cyanogenic glycosides are common plant defensive compounds and occur in different plant families.[23] White clover (*Trifolium repens*, Figure 5.17(c)), a member of the pea family (Fabaceae), has been studied extensively for its content of the glycosides because of its role as a feed plant. Leaves and peel of the tropical passion fruit (*Passiflora edulis*), in the tropical Passiflora family, contain them. Some important food plants contain a good dose of cyanogenic glycosides: examples are lima beans (*Phaseolus lunatus*), the cereal sorghum (*Sorghum* spp.), and cassava roots (*Manihot esculenta*). They all require proper cooking. Many other cereal and legume plants have low doses of cyanogenic glycosides. During the growing period of the plants they have a natural protection against insects and snails because of the cyanogenic glycosides. During the subsequent processing, like milling, steeping, or cooking, the toxins are destroyed as the enzymes from the plant start working on them, and the plant foods become palatable and tasty.

5.11 SOAPY SAPONINS

Some plants or parts of plants produce foaming, soapy mixtures when cut up and mixed with water (Figure 5.19). When lentils or peeled potatoes are soaked in water, bubbles form on the water surface. Agaves or yuccas produce soapy saps that ooze out when cut, especially at the base of their leaf blades. Common plant names, like soap plant (*Chlorogalum pomeridianum*, Figure 5.19(a)), soap tree (*Yucca elata*), or soapwort (*Saponaria* spp.) allude to this characteristic (Figure 5.19(b)), as local native people made use of these plants for natural soaps or shampoos long ago. The compounds that are the cause of these sudsing effects are the *saponins*, their name derived from the Latin word for soap. Saponins are a family of surface-active, mostly bitter plant compounds (Figure 5.20).[24] In plants, saponins provide potent chemical defense against insects and snails.

(a) (b) (c)

Figure 5.19 Saponins in plants.
(a) Soap plant (*Chlorogalum pomeridianum*) with brushy root exposed. (b) Vial with cut-up soap root in water, showing foaming action. (c) Fruits of California buckeye (*Aesculus californica*).

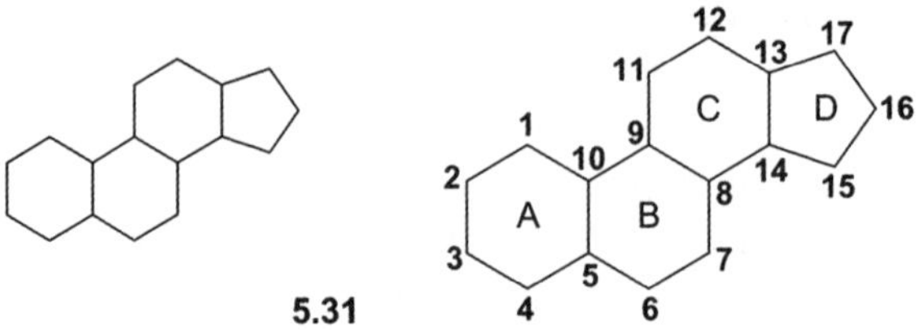

Figure 5.20 Steroid and saponin structures.
The typical four-ring pattern of steroids **5.31** is found in hecogenin **5.32**, the aglycon of a common saponin in agave. C3 usually has a carbohydrate group attached.

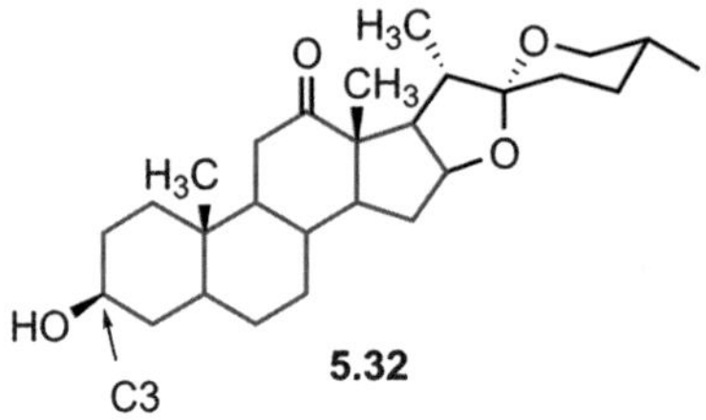

Saponins occur in many different plants and plant families. Their chemical structures are quite diverse, but have certain common features. Chemically they are glycosides of terpenes or steroids. As terpenes, their structures have thirty carbons, *i.e.* they are triterpenes. Instead of the terpene section, they can have a *steroid* ring system. Here we encounter for the first time the typical four-ring structure of *steroids* **5.31** (Figure 5.20). To help in further discussions, the four rings are commonly labeled A – D, and ring carbons are numbered. Hecogenin **5.32** is a common steroidal saponin found in agaves. Its steroid component is highlighted in the structure. Note how additional rings are attached to ring D, with oxygen being part of them. This is typical for saponin structures. Another characteristic of saponins is to have a carbohydrate group bonded to C3 (not shown for greater simplicity). Saponins are glycosides! Not all of the chiral carbons in hecogenin are pointed out in the figure. Try to find the many additional chiral carbons in **5.32**.

Steroids have long been known from animal systems; examples are cholesterol or the steroid hormones. But plants produce steroids, too, like the saponins or the cardiac glycosides in the next section. Biochemically, plant steroids are derived from terpenes (Figure 2.26). Some plant steroids are important sources for steroidal preparations in human medicine as will be shown in the next chapter.

What makes saponins produce soapy mixtures with water? Saponins have both a non-polar and a polar part in their large molecules. This combination of polar and non-polar structural segments is typical of compounds that are soap-like (or "surface-active") in aqueous solutions. The terpene or steroid part of the saponin molecules provides the non-polar, fat-soluble segment. Attached to all saponin molecules is a polar carbohydrate group, making them water-soluble in the plant sap. There are many other, non-saponin structures of compounds in living systems that also exhibit soap-like properties in aqueous solutions. (Artificial soaps that we use for our own cleaning purposes also have a polar and a non-polar part to their molecules.)

Saponins are toxic to insects and to fish, but not to humans in low doses. For that reason some plants that contain saponins have been used by natives to catch fish. Fruits of wild cucumber (*Marah* spp.) and husks of horse chestnut-like buckeyes (*Aesculus californica*, Figure 5.19(c)) are high in saponin content. Native people used to throw the plant parts into creeks or ponds. There, the saponins had a stunning effect on fish as the surface-active saponins interfered with the uptake of oxygen in the animals' gills. The affected fish could then easily be picked out of the water.

5.12 DEFENSIVE CARDIAC GLYCOSIDES

Purple foxglove (*Digitalis purpurea*) (Figure 5.21(a)) has long been known as a poisonous plant, but also as a medicinal plant. The toxic compounds are part of the plant's defense system and include *cardiac glycosides*, also called *cardenolides*. The name of these compounds implies that they affect the activity of the heart muscle. Cardenolides are found in several unrelated plants aside from foxglove, like oleander (*Nerium oleander*, Figure 5.21(b)) and milkweeds (*Asclepias* spp., Figure 5.21(c)). All are plants that figure on toxic plants lists. Cardenolides are poisonous to animals and humans. Despite their toxicity, some of the cardenolides have therapeutic effects and, at appropriate doses, have been used in the treatment of congestive heart failure.

Cardiac glycosides bear a structural resemblance with the steroid saponins shown earlier: they are steroid glycosides. They also have

Figure 5.21 Plants with cardiac glycosides.
(a) Purple foxglove (*Digitalis purpurea*). (b) Oleander (*Nerium oleander*). (c) Showy milkweed (*Asclepias speciosa*), with milky latex showing on leaf.

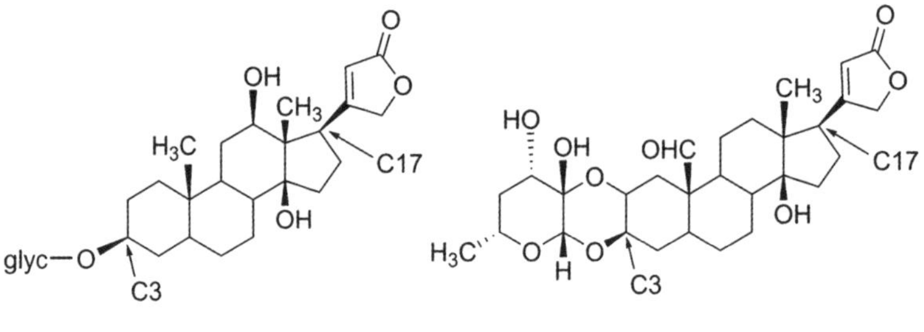

Figure 5.22 Structures of cardiac glycosides.
The typical four-ring pattern of steroids **5.31** is found in digoxin **5.33** from foxglove, and in calotropin **5.34** from milkweeds. Steroid positions C3 and C17 have typical attachments in cardiac glycosides.

foaming, soapy characteristics. Cardenolide structures are distinguished by an interesting, characteristic ring structure attached to ring D in the steroid segment, at carbon numbered 17 (C17) (Figure 5.22). Again polar, carbohydrate-like groups are bonded to C3 in ring A. Note these common structural features in the examples of digoxin **5.33**, a medicinally important cardenolide from *Digitalis* spp. and in calotropin **5.34**, a cardiac glycoside found in milkweeds (*Asclepias* spp.). Again not all of the chiral carbons are pointed out, for greater simplicity.

Some insects are not deterred by the toxins, but have learned to use the plants' defense systems for their own defense. Read more about these fascinating plant-insect interactions in Box 5.2.

BOX 5.2 Insects and Plant Defenses

Some types of aphids infest purple foxglove without suffering any apparent harm from the plant toxins.[25] Similarly, other aphids feed on oleanders and milkweeds without being deterred by the cardenolides, but seem to obtain protection of their own through the poisons. A famous insect-cardenolide interaction involves the monarch butterfly (*Danaus plexippus*, Figure 5.23). Its caterpillars are distinctly patterned (Figure 5.23(a)) and feed on the milky saps of milkweeds without being adversely affected by the toxins. The emerging butterflies, also brightly colored (Figure 5.23(b)), still contain the milkweed toxins and store them in their bodies without harm, yet make birds sick that prey on them. Milkweed beetles (*Tetraopes tetrophthalmus*) also use the plant toxins for their own defense. It is remarkable how many insects that contain toxins have dramatic coloring: orange-black for the monarch butterfly and the milkweed beetle, bright yellow for the oleander aphid, and colorfully striped for the caterpillar

Figure 5.23 Insects using plant defenses.
(a) Caterpillar of monarch butterfly (*Danaus plexippus*) feeding on a milkweed seed pod. (b) Monarch butterfly on flowers of orange milkweed (*Asclepias tuberosa*). (Photos by Tellur Fenner.)

of the monarch. These warning colors may well alert predators to the toxicity of the intended prey.

The above examples are just a few among the numerous insect-plant interactions where insects have learned to use the chemical defenses of plants for their own defense, as part of the coevolution of plants and insects.

5.13 POTENT ALKALOIDS

The best-known chemical plant defenses are likely the plant bases known as alkaloids, probably because so many of them have distinct effects on the human body, be it as stimulants, as potent toxins, or as addictive substances. Caffeine, nicotine, morphine, and quinine are examples of alkaloids. Plants produce thousands of different alkaloid structures.[26] It is interesting to note that some plant families are especially rich in alkaloids, like the nightshade family (Solanaceae) or the pea family (Fabaceae). Alkaloid names tend to end in "-ine", and their plant origin is often reflected in the name of the compound, like solanine from potato plants (*Solanum tuberosum*), coniine from the poison hemlock plant (*Conium maculatum*, Figure 5.24(a)), or nicotine from the tobacco plant (*Nicotiana* spp., Figure 5.24(b)). Earlier plant pictures in this book showed deadly nightshade (*Atropa belladonna*, Figure 1.1(c)) that contains the toxic alkaloid atropine, and a coffee plant (*Coffea* spp., Figure 1.22(b)), a source of caffeine.

Figure 5.24 Plants with alkaloids.
(a) Poison hemlock (*Conium maculatum*). (b) Tobacco plant (*Nicotiana* sp.). (c) Jimson weed or thorn-apple (*Datura wrightii*).

Poisons and Other Defenses 141

Alkaloids are a very diverse group of compounds. As the following descriptions show there is some flexibility in the definition of what an alkaloid is. All are secondary metabolites that have nitrogen in their organic structures. They are usually alkaline or basic (thus their name "alkaloids" or "plant bases"). Basic alkaloids are found in plants as salts which makes them soluble and transportable in the aqueous plant saps. Most alkaloids have complex ring structures, and chemists frequently group them according to their ring systems. Alkaloids are mostly found in higher plants—although some are known to occur in animals, like salamanders and poison arrow frogs. Certain fungi, such as ergot, a fungus growing on rye, also contain alkaloids. A bitter taste is a common property of alkaloids.

To illustrate alkaloid structures (Figure 5.25), we begin with the relatively simple structure of coniine **5.35**, a toxic alkaloid found in the poison hemlock plant. This plant and its alkaloid are of ill fame as the Greek philosopher Socrates was sentenced to death by drinking an extract of poison hemlock. Note that the molecule of coniine has a chiral center. In hemlock plants only one of the enantiomers is found. The alkaloid nicotine **5.36** is found in all plants of the genus *Nicotiana* (tobacco plant). It is a potent

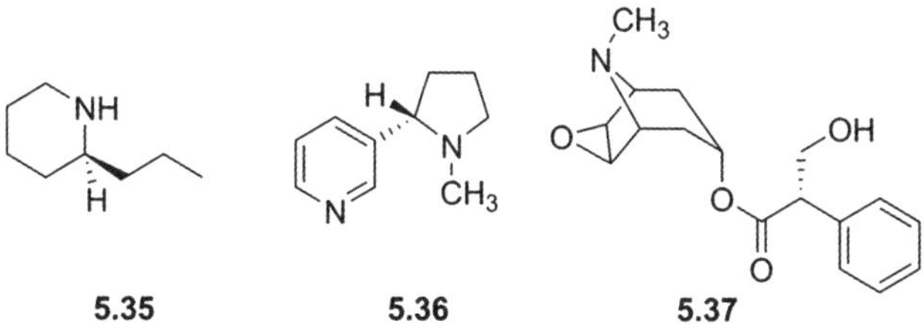

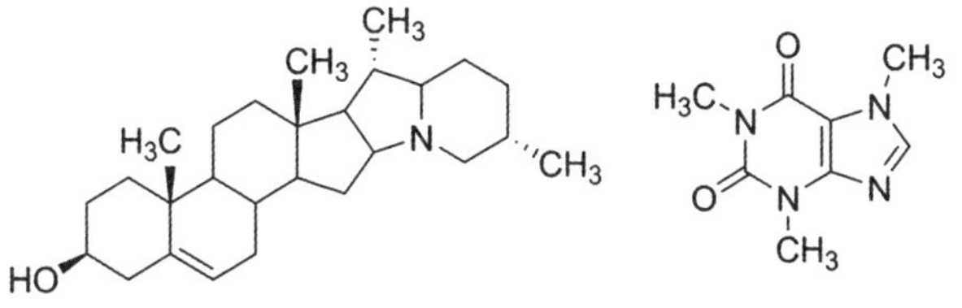

Figure 5.25 Structures of some plant alkaloids.
Coniine **5.35** is found in poison hemlock, nicotine **5.36** in tobacco plants, and scopolamine **5.37** in *Datura* spp. Solanine **5.38** occurs in greening potatoes, and caffeine **5.39** in coffee plants.

insecticide and a highly toxic compound to humans, too. Scopolamine **5.37** is an alkaloid in jimson weed or thorn-apple (*Datura* spp., Figure 5.24(c)). It has a complex structure with several different rings. While highly toxic, it can be used in small doses to counteract motion sickness. Solanine **5.38** is a toxic alkaloid with a steroid structure. It is a saponin, too. Solanine forms when potatoes are greening; it is also found in their sprouts. As for complex ring structures, review the molecule of quinine **5.19**, an intensely bitter alkaloid. Caffeine **5.39** is the stimulant in coffee and black tea, also in small quantities in cacao. Pure caffeine has a bitter taste. We will encounter alkaloids again in the next chapter, especially as medicines and as psychoactive plant compounds.

5.14 CONCLUSION

The extent of this chapter reflects the great chemical diversity of plant defenses and also their many modes of action. Yet, there are aspects of chemistry that connect the classes of defensive compounds.

As we contemplate the chemical families that provide the defenses, we encounter terpenes and phenolics as recurring themes. We find monoterpenes as repelling smells. The small sizes of their molecules make them volatile. Larger terpenes, with fifteen and twenty carbon atoms, compose resins and some plant irritants. Even larger terpene molecules, with thirty carbon atoms, compose the backbone of some of the surface-active saponins. Finally, terpenes as polymers form natural rubber. Phenolics contribute to many resinous mixtures and often contribute distinct, sharp smells. Phenolic groups are part of the large structures of tannins and give them their acidic character.

This leads us to acids and bases among the plant defenses. Acidic compounds are represented by phenolic groups and carboxylic acids. The latter provide sour tastes in many plants. Defensive compounds with basic properties are represented by the large family of alkaloids. We encountered also the case where sharp, water-insoluble calcium salts of oxalic acid form a type of physical defense in leaves or roots.

Several classes of defensive plant compounds develop only once an animal starts feeding on leaves or other plant parts. At that time, enzymes from neighboring plant cells get into action and break apart compounds. As a result, volatile compounds are released that have a sharp or otherwise repelling smell or that are downright toxic. This type of mechanism is advantageous for the plants as they can conserve the metabolites unless they are attacked. We encountered the repelling smells from glucosinolates and the sulfurous smells from *Allium* species that required the interaction with enzymes for their formation. Cyanogenic glycosides that produce toxic hydrogen cyanide also need enzyme action to release the defensive compounds.

With the class of saponins, the typical properties of surface active compounds were introduced, *i.e.* of compounds that act like soap in water. Their relatively large molecules have polar and nonpolar segments. The section on saponins also introduced the characteristic four-ring structure of steroids. It is a structure that is found in many natural products, in animals as well as in plants. We will encounter it again in medicines from plants in the next chapter.

This chapter frequently alluded to human uses of plant defensive compounds, *e.g.* as medicines, as food components, or as natural preservatives. At this point, we have built a considerable basic knowledge of organic chemistry as it relates to plants. We are going to apply this knowledge in the next chapter that describes in more detail some of the human uses of plant compounds.

REFERENCES

1. P. R. Cheeke, *Natural Toxicants in Feeds, Forages, and Poisonous Plants*, Interstate Publishers, Danville, IL, 1998.
2. M. B. Isman, in *Agricultural Applications in Green Chemistry*, ed. W. M. Nelson, ACS Symposium Series, American Chemical Society, Washington, DC, 2004, **887**, ch. 4, pp. 41–51.
3. E. Breitmaier, *Terpenes: Flavors, Fragrances, Pharmaca, Pheromones*, Wiley-VCH Verlag, Weinheim, 2006.
4. J. R. Hanson, *Chemistry in the Garden*, The Royal Society of Chemistry, Cambridge, 2007.
5. J. H. Langenheim, *Plant Resins*, Timber Press, Portland, OR, 2003.
6. A. Wei and T. Shibamoto, *J. Agric. Food Chem.*, 2010, **58**, 7218.
7. J. B. Lambert and G. O. Poinar, *Acc. Chem. Res.*, 2002, **35** (8), 628.

8. M. Freemantle, *Chem. Eng. News*, 2007, **85** (11), 41.
9. R. T. Cardé and W. J. Bell, *Chemical Ecology of Insects 2*, Chapman & Hall, New York, NY, 1995.
10. E. Block, *Garlic and Other Alliums*, The Royal Society of Chemistry, Cambridge, 2010.
11. P. Atkins, *Atkins' Molecules*, Cambridge University Press, Cambridge, 2nd edn, 2003.
12. T. Betsche and B. Fretzdorff, *J. Agric. Food Chem.*, 2005, **53**, 9751.
13. V. R. Franceschi and H. T. Horner, Jr., *Bot. Rev.*, 1980, **46**, 361.
14. H.-D. Belitz, W. Grosch and P. Schieberle, *Food Chemistry*, Springer-Verlag, Berlin, 3rd edn, 2004.
15. G. B. Kauffman and R. B. Seymour, *J. Chem. Educ.*, 1990, **67**, 422.
16. D. F. Hegerhorst, D. J. Weber, E. D. McArthur and A. J. Khan, *Biochem. Syst. Ecol.*, 1987, **15**, 201.
17. J.-L. Giner, J. D. Berkowitz and T. Andersson, *J. Nat. Prod.*, 2000, **63**, 267.
18. H. Schildknecht, *Angew. Chem., Int. Ed.*, 1981, **20**, 164.
19. G. Goel, H. P. S. Makkar and G. Francis, *Int. J. Toxicol.*, 2007, **26**, 279.
20. E. L. Rice, *Allelopathy*, Academic Press, Orlando, FL, 2nd edn, 1984.
21. C. H. Muller, W. H. Muller and B. L. Haines, *Science*, 1964, **143**, 471.
22. J. B. Harborne, *Introduction to Ecological Biochemistry*, Academic Press, London, 4th edn, 1993.
23. M. Zagrobelny, S. Bak, A. Vinther Rasmussen, B. Jørgensen, C. M. Naumann and B. Lindberg Møller, *Phytochemistry*, 2004, **65**, 293.
24. J.-P. Vincken, L. Heng, A. de Groot and H. Gruppen, *Phytochemistry*, 2007, **68**, 275.
25. P. Jolivet, *Interrelationship Between Insects and Plants*, CRC Press, Boca Raton, FL, 1998.
26. M. Hesse, *Alkaloids*, VHCA, Zürich, and Wiley-VCH, Weinheim, 2002.

CHAPTER 6
Plants and People

6.1 INTRODUCTION

This final chapter provides an illustration of the vital and important roles that plants have in our lives. Above all, plants provide oxygen to breathe. Equally important is their role as the source of essential sugars, fats, and amino acids, and of vitamins, too. If we like to add flavorful herbs and spices to our food, we obtain them from plant sources. Plants were the origins of most medicines, and many medicinal plants and their active compounds are still in use. Moreover, a large number of modern pharmaceuticals have been derived from plant compounds, and plants continue to be a source of the search for new drugs. Psychoactive plants have been used by humans since ancient times. Plants also provide fibers for clothing, and plant materials were the origins of dyes to color the fibers. Furthermore, the perfume and cosmetics industry would not exist without scents and oils from plants.

Plants have many other important uses. For thousands of years, plants have provided construction materials, in the form of wood, to build houses and boats. Wood also provides heat for cooking and warmth. As for other sources of energy, fossil plants are the origins of crude oil and coal, and ancient plant materials have contributed to the formation of natural gas through anaerobic decomposition.

The Chemistry of Plants: Perfumes, Pigments, and Poisons
Margareta Séquin
© Margareta Séquin 2012
Published by the Royal Society of Chemistry, www.rsc.org

Much history is connected with the way people have used plants through the ages, and some of these uses and how they led to contemporary applications will be described. Ancient trading included many plant materials, like spices, medicines, dyes, and drugs. Major plant crops still impact the economy in today's world markets. Chemical processes to make plant compounds useful or appealing for human use have long traditions. (Just think of the fermentation of grapes!). In more recent times, the chemical compositions of plant compounds have provided inspiration for the development of many new compounds, namely in the field of pharmaceuticals. In recent decades, research has made it possible to alter the genetic material of plants, in order to obtain crops with desirable properties. An introduction to genetically modified plants concludes the chapter.

The following sections are not intended to be—nor can they be—a comprehensive treatise on people's uses of plants. Instead, the topics describe select important human uses of plants and connect them with the organic compounds that are responsible for their usefulness. Themes of organic chemistry that were addressed in the previous chapters will be revisited, and a few new concepts will be introduced. Therefore, this chapter rounds off our introduction to organic chemistry as it relates to plants.

6.2 FOODS FROM PLANTS

6.2.1 Essential Primary Metabolites

"Natural phytonutrients: power from Nature!" proclaimed an advertising line on a package of mixed vegetables ("phyto" being Greek for plant). Indeed, plants can provide all the nutrients that humans require (Figure 6.1).[1-3] Plant products supply the basic metabolites that we need to obtain energy and to metabolize into other vital compounds. The following summary of primary metabolites includes examples of plants that can provide them. You may want to refer to Chapter 2 for a reminder of chemical structures and properties of primary metabolites.

Carbohydrates are essential components of human nutrition. They provide energy and have central roles in the metabolisms of animals and plants. Green plants can produce simple sugars, through photosynthesis from water and carbon dioxide. From the

Figure 6.1 Foods from Plants.
Market displays of (a) mixed vegetables, (b) beans, and (c) grains.
(6.1(b): photo by Liselotte Wespe.)

basic sugars, plants synthesize starch and all other organic compounds. Animal systems have the necessary enzymes to metabolize starch into simpler sugars and then can use them as sources of energy. Not surprisingly, plants that are rich in starch, like cereals, rice, potatoes, and corn, are important human staple foods. Nondigestible carbohydrates from plants, like cellulose, provide needed fiber in human nutrition.

Fats and oils are another major source of energy from nutrition. Recall that fats and oils are esters of fatty acids and glycerol. Essential fatty acids are required in our nutrition but cannot be synthesized by human and animal metabolisms; they need to be obtained through plant-related foods. Linoleic acid **2.14** (Figure 2.13) is an example of an unsaturated fatty acid that is essential and that plants synthesize from carbohydrates. Oils from plant seeds, like peanuts, soybeans, sunflower seeds, or coconuts, provide essential fatty acids. Recall how plant oils tend to be highly unsaturated and therefore are liquid at room temperature (Chapter 2.4). An exception is palm oil, extracted from the pulp of the fruit of the oil palm (*Elaeis guineensis*, Figure 6.2(c)). This mostly saturated oil is semi-solid at room temperature. It is also more stable towards oxidation than highly unsaturated plant oils. Remember also that natural oils from plant sources have *cis* orientation around their double bonds (Figure 2.14). Unnatural *trans* fats, with a *trans* orientation around double bonds, are obtained when oils from natural sources are treated with hydrogen gas. This process is useful in producing more solid fats, as in making margarine from vegetable oils. In partial hydrogenation, hydrogen atoms add to the carbon-carbon double bonds in the triglyceride molecules and transform some of the double bonds into single bonds. This produces more saturated fats that tend to be solid at room temperature. But these reactions can also lead to the formation of *trans* fats that are undesirable in nutrition because of their connection with raising levels of LDL ("bad") cholesterol in blood.

Many types of plant seeds serve as sources of basic nutrients for humans: cereals, corn or maize, rice, beans, and nuts are all seeds. This is quite understandable as plant seeds have to provide nutrients for future plants, too. Plant seeds not only supply starch and oils, but also amino acids and proteins. Eight of the twenty amino acids that compose vital proteins in the human body must be obtained through our diet, and plant sources can provide them. If you check back in Chapter 2, you find the structures of some amino acids in Figure 2.19. In a mostly vegetarian diet, care has to be taken to obtain adequate amounts of the essential amino acids because not all plants can provide a sufficient supply of them. Methionine **2.20** and lysine **2.22** are two essential amino acids. Rice, wheat, and corn are low in lysine content, but provide adequate amounts of the other amino acids. Legumes, like beans, on the other hand, are good sources of lysine but are deficient in methionine and a few other essential amino acids. It is no coincidence that many vegetarian and traditional dishes are based on combinations of rice and soy, or corn and beans. These combinations provide the full complement of required amino acids. Many plant breeding programs work on producing crop plants like maize that have all the essential amino acids in adequate supply.

In addition to the primary metabolites, the human diet must, of course, include water and specific metals in the form of ions. Vitamins must also be a part of our nutrition, and plants can provide them (Figure 6.2). These vital compounds are described next.

6.2.2 Vitamins

Our daily diet requires the intake of some organic compounds that are needed only in very small amounts, yet that are essential for health. They are the *vitamins*, so named because they are vital in human nutrition and because the first compound discovered to have this role happened to be an amine. It was vitamin B_1 or thiamine **6.1**, a vitamin found in the bran and germ of brown rice (*Oryza sativa*). Because the oils in the germs of rice kernels are unsaturated oils, they turn rancid rather quickly. To avoid this spoilage, rice milling machines were developed in the 1800's that stripped the seeds from bran and germs. The product was white rice that is less perishable, but has a very low thiamine content. As a consequence, people who lived mostly on white rice developed beriberi, a disease that affects the nervous system of humans. It is the result of a deficiency in vitamin B_1.

Vitamins have a wide range of organic structures (Figures 6.3 and 6.4). They are usually classified either as water-soluble or as fat-soluble. Plants can provide all of the vitamins needed for

Figure 6.2 Plants and vitamins.
(a) Tangerines (*Citrus* spp.), a source of vitamin C (ascorbic acid). (b) Carrots, source of β-carotene (pro-vitamin A). (c) Fruits of oil palm (*Elaeis guineensis*), source of vitamin E (tocopherol). (Photo from Wikimedia Commons public domain.)

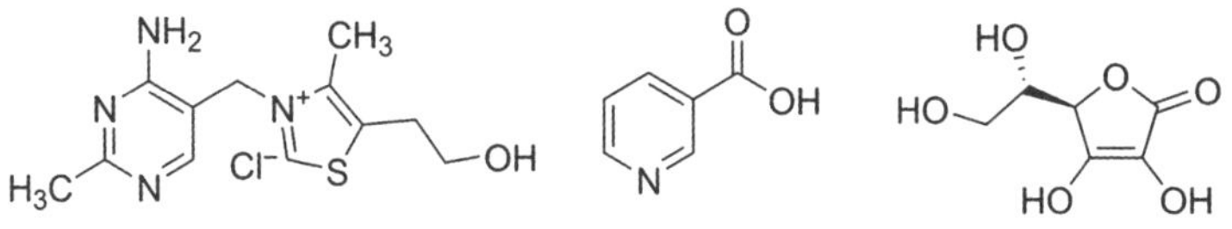

Figure 6.3 Water-soluble vitamins.
Thiamine (vitamin B_1) **6.1**, niacin (vitamin B_3) **6.2**, and ascorbic acid (vitamin C) **6.3** are examples of water-soluble vitamins.

human nutrition. But vitamins are not only essential for human health. They have important roles in plant metabolism.[4] Many act as *cofactors*, *i.e.* as compounds that are needed to activate enzymes. In addition, many vitamins are powerful antioxidants. We are next going to look at some representative vitamins and their structures, together with their roles in plant metabolism and in human nutrition.

Vitamins in the B group, of which thiamine (vitamin B_1) is an example, are all water-soluble (Figure 6.3). Many of them have antioxidant activity. They are mostly found in seeds. The polar structure of thiamine **6.1**, together with an ionic site in the molecule, points to its water-solubility. Interestingly, thiamine contains sulfur in its structure (therefore the name *thi*amine). Another water-soluble vitamin of this group is niacin (vitamin B_3), also known as nicotinic acid **6.2**. Niacin is involved in the synthesis of NAD (nicotinamide adenine dinucleotide) which we encountered in the photosynthesis reaction steps (Table 1.2). It has a key role in cellular oxidation-reduction chemistry. Vitamin B_3 deficiency in humans causes the disease pellagra. There is an interesting chemical story that connects corn with niacin. Corn or maize contains a relatively low amount of this vitamin, and the niacin that is present is not very bioavailable. In a process from Mesoamerica, dating back to more than 1000 BC, maize or other grains are soaked and cooked in alkaline solutions, followed by washing and rinsing and then grinding into flour. This process breaks down plant cell walls and makes niacin more readily available (and also destroys microorganisms). The flour is called "masa" and is used for making tortillas. It is always amazing how people figured out these processes that brought them better health.

Among the water-soluble vitamins, ascorbic acid (vitamin C) **6.3** is abundantly present in plants, especially in fruits of citrus plants (Figure 6.1(a)) and in green leaves. A deficiency in vitamin C causes scurvy in humans, a disease that is cured by eating citrus fruit and green vegetables. The many OH groups in molecules of ascorbic acid make vitamin C a polar, highly water-soluble compound. Note also that ascorbic acid is an asymmetric compound with a chiral carbon. Vitamin C is a potent antioxidant.

A more detailed explanation of the nature of antioxidants makes us understand why they have such important roles in plants and in animal systems. *Antioxidants* are molecules that inhibit the (undesirable) oxidation of other molecules. Normal metabolism of oxygen forms *reactive oxygen species (ROS)* as byproducts. ROS are highly reactive and when in abundance destroy vital molecules. Examples of ROS include peroxides. ROS are so aggressive because they are *radicals*, *i.e.* atoms or molecules that have unpaired valence shell electrons. Recall how electrons strive to form pairs when bonding. Unpaired electrons are unstable which makes radicals seek electrons to pair with. Radicals can form as a result of high heat or of irradiation with high energy light. Many radicals have important roles in cell signaling in living organisms. However, during times of environmental stress (*e.g.*, due to high ultra violet or heat exposure), ROS levels can increase dramatically. This may cause significant damage to cell structures because the radicals, seeking electrons to pair with, can break bonds and thus damage macromolecules like DNA and proteins. Antioxidants, like vitamin C, counteract this stress by reacting themselves with ROS. The antioxidants are destroyed in the process, but inactivate the ROS before vital molecules are damaged. Antioxidants in foods and their modes of action are a field of vigorous research in food science.[5]

Vitamin A (retinol) and vitamin E (α-tocopherol) are examples of fat-soluble vitamins (Figure 6.4). We encountered β-carotene **6.4** earlier, as a plant pigment in Chapter 4. It appears here as provitamin A, as it is the main biosynthetic precursor for vitamin A **6.5**. Humans can rapidly metabolize β-carotene into this vitamin. Other carotenoids can also serve as provitamins, but to a lesser extent.

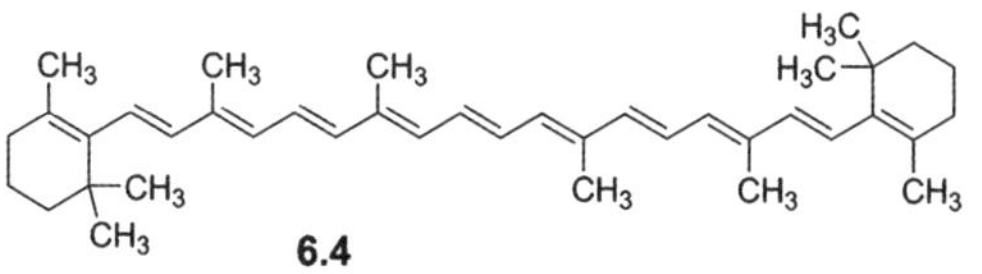

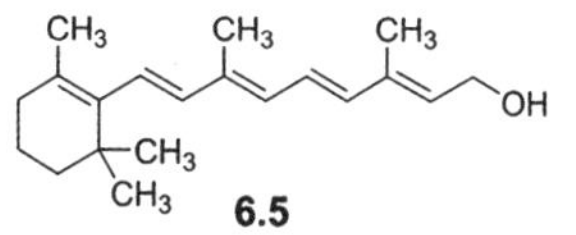

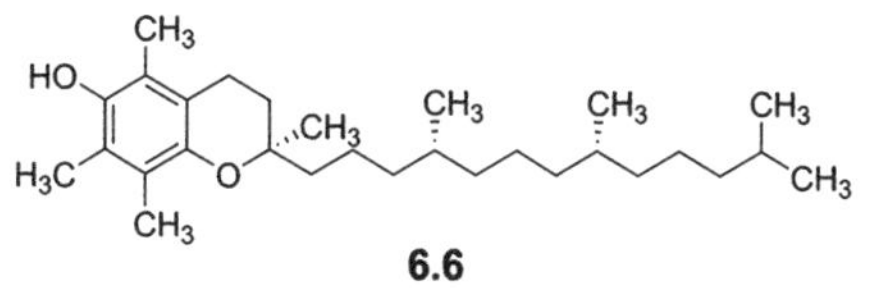

Figure 6.4 Fat-soluble vitamins.
β-Carotene (provitamin A) **6.4**, vitamin A (retinol) **6.5**, and α-tocopherol (vitamin E) **6.6** are insoluble in water.

Vitamin A is essential in the vision process in humans. A lack of vitamin A leads to night blindness. Carotenoids are present in all photosynthetic organisms. β-Carotene accumulates especially in leaves where it has an important protective role as antioxidant, counteracting an excess of ROS formed during photosynthesis. Carrot roots (Figure 6.2(b)) have even higher carotene concentrations than green leaf vegetables. The hydrocarbon structures of β-carotene and vitamin A make them fat-soluble. Notice also their terpene structures.

The long hydrocarbon sidechain of vitamin E (α-tocopherol) also has an isoprenoid structure. Vitamin E is especially found in seeds. Oils from these seeds are desirable sources of vitamin E in human nutrition. An example is palm oil, from the fruits of the tropical oil palm (Figure 6.2(c)). Vitamin E is a strong antioxidant that protects the plant seeds.

Solubility differences among vitamins have significant implications for human nutrition and health. Fat-soluble vitamins, like vitamins A and E, are stored in lipid cells and can accumulate. While too low an intake of a vitamin leads to health problems, ingesting excess amounts has detrimental effects on human health as well. Too high a dose is not much of a problem in the case of water-soluble vitamins as they are excreted by urine. But if excessive doses of fat-soluble vitamins build up in the human body, they have damaging consequences.

6.2.3 Flavors, Herbs, and Spices

The addition of flavorful herbs and spices to foods has been a tradition in most cultures of the world. Plants and plant parts with distinct and often intense flavors are used either dried or in fresh form as seasonings or spices.[6] Herbs are usually dried or fresh green leaves and stems from aromatic plants that you might find in your garden. Spice plants, on the other hand, are often from tropical countries, and spices come from their roots, seeds, fruits, flower buds, or even their bark. Black pepper (from *Piper nigrum*) and nutmeg (*Myristica fragrans*, Figure 6.5(a)) are examples of seeds used as spices. The bark of the cinnamon tree (*Cinnamomum aromaticum*) is the source of cinnamon, and the flower buds of *Syzygium aromaticum* supply the spice cloves (Figure 6.5(a)). The stigmas of the saffron crocus (*Crocus sativus*) are the sources of

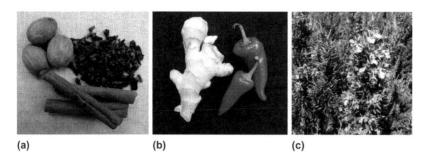

Figure 6.5 Spices and herbs.
(a) Nutmeg, cloves, and cinnamon. (b) Ginger root and red chili peppers. (c) The herb rosemary.

saffron. Turmeric (*Curcuma longa*) and ginger (*Zingiber officinale*) (Figure 6.5(b)) are examples of spices from roots. Hot tastes are supplied by the fruits of paprika and chili peppers (*Capsicum annuum*, Figure 6.5(b)). Leafy herbs, like parsley, sage, thyme, oregano, and rosemary (*Rosmarinus officinalis*, Figure 6.5(c)), have been used to flavor foods since ancient times. A large number of herbs originate from Mediterranean climates. Their flavorful compounds as well as those that compose spices have mostly defensive functions in plants, be it to repel insects with their strong scents or to deter browsers from eating the plants.

Much history is tied to the trade of spices.[7] Their exotic flavors—and the fortunes that their sales brought—were the draw for explorations and voyages by European powers to the Indies during the Middle Ages, and wars were fought to obtain monopoly of the spice trades. The name "Spice Islands" (now the Moluccas) refers to a group of islands in Indonesia, west of New Guinea, where nutmeg and cloves originated.

Many herbs and spices were originally used for their medicinal properties, like sage, or ginger, and were later added for their tastes, too. Some of the flavorful plant products have been used as food preservatives, like nutmeg, as they are strong antioxidants. In the past, herbs and spices were also used to cover the taste of spoiled foods. Gradually the flavors found entry into mainstream cooking. Nowadays we enjoy the flavors of herbs and spices, but are also learning about their health benefits as antioxidants.[8]

Spice and herb flavors are always mixtures. Their composition is affected by the age of the plants at time of harvest as well as by the particular species of plant. Often one compound stands out and characterizes a spice or herb. Taste, flavor, and odor all combine in the olfactory experience. Earlier we encountered vanillin **3.5**, the world's most popular flavor, from the seedpods of vanilla orchids. The odor of cinnamon is characterized by cinnamaldehyde **6.7** (Figure 6.6). Saffron, one of the oldest and most expensive spices, owes its taste mainly to picrocrocin **6.8**, a monoterpene bonded to glucose. (Can you find the isoprene units?). The glycoside structure of picrocrocin, a bitter-tasting compound, makes it soluble in water. Some of the compounds that were shown in earlier chapters as attractive or defensive plant smells reappear here as components of flavors in herbs and spices. Quite commonly, these compounds contribute to the flavors of several different spices and herbs. Eugenol **3.6** is the major flavor component in cloves, but is also found in allspice. α-Pinene **5.3** is part of the flavor composition of pepper, nutmeg, and rosemary. Most components of flavors

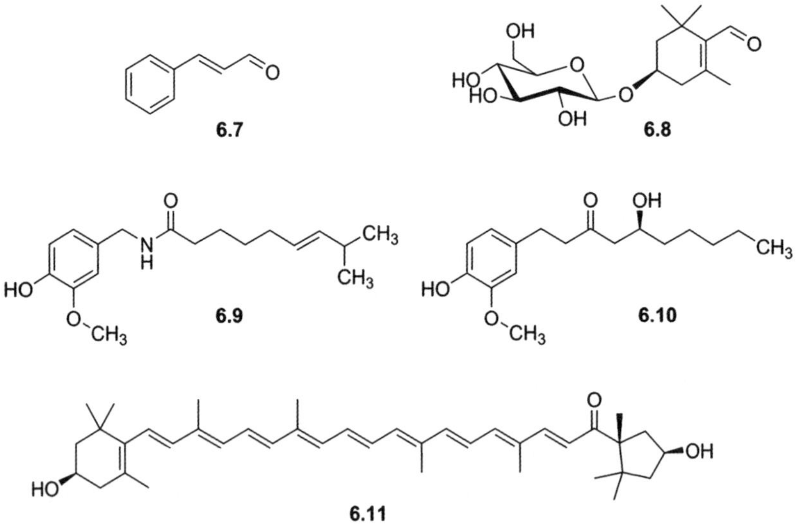

Figure 6.6 Compounds in spices.
Cinnamaldehyde **6.7**, and picrocrocin **6.8** are characteristic flavor components in cinnamon and saffron, respectively. Capsaicin **6.9** and gingerol **6.10** provide the pungent flavors to peppers and ginger roots, respectively. The red pigment of paprika is capsanthin **6.11**.

are either aromatic compounds (like vanillin, cinnamaldehyde, and eugenol) or have monoterpene structures (like picrocrocin or α-pinene). They are quite volatile, and most are fat-soluble.

Some compounds in spices cause a burning sensation. They have somewhat different structures compared to the earlier examples. The compound that causes the hot reaction in paprika and chili peppers (*Capsicum annuum*) is capsaicin **6.9** (Figure 6.6). The pungent principle of ginger roots is gingerol **6.10**. There are similarities in the two structures: both are larger molecules than the flavors seen before and have phenolic structures. They are non-volatile. Some spices not only add flavor but also color to foods. The compounds that provide the aromas and those responsible for the color are not the same ones. Colorful pigments must have the pattern of conjugated double bonds in their structures. You find this pattern in the molecule of capsanthin **6.11**, the red carotenoid pigment in paprika and chili peppers. Saffron adds yellow coloring to food. Its structure and use as a yellow dye will be addressed later in this chapter.

When enjoying a meal, we can reflect on how many of its components have been provided by plants. Vegetables, of course, breads, rice and pasta, and flavorings from herbs and spices all have plant sources. To round off our plant-derived meal, we can add beverages, like wine or beer, or coffee and tea, and some dark chocolate for dessert.

6.3 PLANT MEDICINES

6.3.1 History and Introduction

Healing plants have helped humans alleviate pains and ailments since the early beginnings of mankind.[9–11] Egyptian scrolls from around 1500 BC mention medicinal herbs. An extensive Chinese medicinal book, dating back to the 1st century AD, classified numerous types of plants and their useful parts, together with some animal materials; the book was compiled from even earlier experiences. The ancient Greeks and Romans, and native tribes of the Americas all used medicinal plants. During the middle ages, herb gardens were an important part of medieval monasteries, and medicinal gardens are still a tradition in many public and private gardens (Figure 6.7(a)). Through trial and error, people learned

Figure 6.7 Medicinal plants.
(a) Medicinal herb garden. (Photo by Ruth Marent.) (b) Market display of medicinal herbs for sale. (Photo by Joan Hamilton.)

which plants were effective in treating diseases or in soothing pain. Sometimes the reaction of animals towards plants led people to further investigate the plant sources. The acquired knowledge was continued through the generations through oral and written traditions. In many parts of the world, plants still provide most medicines, and traditional markets usually have displays of herbal medicines (Figure 6.7(b)).

Many common names of medicinal plants refer to their uses, like self-heal (*Prunella vulgaris*) or feverfew (*Tanacetum parthenium*). Systematic names of medicinal plants frequently carry the species name *officinalis* or *officinale*, meaning "used in a pharmacy" or "used in medicine", like *Salvia officinalis* for a sage, *Taraxacum officinale* for dandelion, or *Zingiber officinale* for ginger.

Medicinal plants always contain mixtures of potentially active compounds, and their compositions are affected by the origins of plants and their time of harvest. In general, medicinal herbs are more easily accessible for people than prescription drugs and less expensive, but the composition and concentration of their active compounds vary. Additional compounds with pharmacological activity may be present, and their added presence may have beneficial effects or unwanted side reactions. In contrast, contemporary Western scientific medicine prefers defined single compounds that can be tested and prescribed in exact doses. Nevertheless, many modern pharmaceuticals have direct connections with plants. Some extracts from plants, obtained by steeping plant materials in solvents like water or alcohol to extract the active compounds, are commonly used in Western medicine. An example is clove oil, used as a topical anesthetic in dentistry. Some modern pharmaceuticals, like quinine or morphine, are still obtained from their plant sources as a complete synthesis from simple organic compounds is too complex and expensive.

The isolation of pharmacologically active plant compounds is a lengthy process! Plants known for their traditional uses often serve as sources. Extractions of suitable plant materials usually yield complex mixtures that have to be separated into their components. Various, sometimes tedious separation methods are required to obtain the pure active compounds. Commonly, large amounts of plant material are needed to obtain just a few grams or even milligrams of the desired substances. After successful isolation and purification of the compounds, the chemical structures have to be determined. Most pharmacologically active compounds have complex structures, commonly with several chiral centers. Only modern instrumentation has made it possible to pinpoint their exact structures. As a further step in the procedures, the physiological activities of the compounds have to be tested. The successful isolation and structure elucidation of pharmaceuticals from plants to their implementation as pharmaceuticals takes years, as illustrated with the historic examples in the next section.

Plants have evolved a huge array of highly diverse structures in their defense against organisms that threaten them. Many of these compounds have proved beneficial in our own struggle with pathological organisms. Their chemical structures are as diverse as plant defensive compounds, but alkaloids and steroids figure prominently. Read next about the discovery of some established pharmaceuticals from plants that are still in use nowadays.

6.3.2 Classic Medicines from Plants

Paracelsus, a medieval physician and alchemist of the early 15th century, was first to promote the idea "Dosis sola facit venenum" or "the dose alone makes the poison". He studied the possibility of using very small amounts of toxic mercury and of opium as healing agents. Only much later, in the 19th century, was the isolation of active plant compounds accomplished as methods of modern chemistry became available.

At the beginning of the 19th century, Sertürner, a German pharmacist, isolated morphine and codeine from opium, the dried latex from the opium poppy (*Papaver somniferum*, Figure 6.8(a)). The complex chemical structure of morphine **6.12** (Figure 6.9) was figured out only in 1925, by British chemist Sir Robert Robinson. In 1952, the synthesis of morphine was accomplished from simple

Figure 6.8 Plant sources of classic medicines.
(a) Seed pod of opium poppy (*Papaver somniferum*), slit to show latex. (b) Cinchona tree with bark detail (*Cinchona pubescens*). (c) Ripe fruit of deadly nightshade (*Atropa belladonna*).

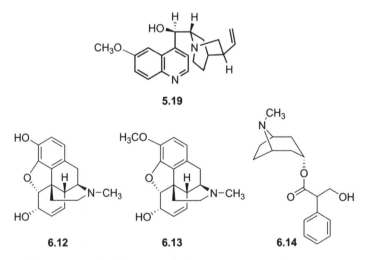

Figure 6.9 Structures of classic medicinal plant compounds.
Quinine **5.19**, morphine **6.12**, codeine **6.13**, and atropine **6.14** are all alkaloids.

Plants and People 159

organic compounds. Morphine is still one of the strongest painkillers known. It is also an addictive drug. Codeine **6.13** is another *opiate*, *i.e.*, an alkaloid found in the latex of opium poppy and a component of opium. The structure of codeine is very similar to morphine, yet the compound is a much weaker painkiller. On the other hand, it is used as a cough suppressant. This illustrates how only minor changes in the chemical structure of a compound can lead to major changes in its pharmacological activity.

Much admiration is due to early scientists who, in spite of limited instrumentation, were able to figure out many of the structural components of complex medicinal compounds. The final proof of the correct assignment was usually accomplished much later by performing a *total synthesis*, *i.e.*, a synthesis of the compound of interest from simple, easily available organic compounds. It is one of the ultimate challenges for chemists to design a sequence of chemical methods that lead to the synthesis of a complex natural compound. Several Nobel prizes have been awarded for the successful total synthesis of plant compounds with intricate structures. Many of the endeavors to find an efficient synthesis have also led to new discoveries as shown in the next example.

The complex structure of quinine **5.19** was introduced in Chapter 5. (The structure is shown again in Figure 6.9.). It is the very bitter alkaloid from the bark of the tropical Cinchona tree (*Cinchona pubescens*, Figure 6.8(b)). Quinine is still used as an important antimalarial.[12] Its isolation from the plant was successfully accomplished during the 19th century, by French chemists Pelletier and Caventou. In 1856, Sir William Henry Perkin, an English chemist, tried the synthesis of quinine as a student—and in the process made the first synthetic dye instead (see section 6.5.4)! The actual synthesis of quinine was accomplished only in 1944, by American chemists Woodward and Doering. Robert Woodward was awarded a Nobel Prize in 1965 for his successful syntheses of many natural products with intricate structures, quinine being one of them. The many synthetic steps to synthesize quinine are especially challenging as the compound, like so many other plant molecules, has several chiral centers, all of which need to be correctly oriented for proper activity.

Another alkaloid from plants that is still in use in modern medicine is atropine **6.14** (Figure 6.9), one of the toxic alkaloids from the deadly nightshade plant (*Atropa belladonna*, Figure 6.8(c)).

A measure used for the toxicity of a compound is the LD_{50} or lethal dose that kills 50 % of a population of test animals, usually mice or rats. The LD_{50} for atropine is 750 mg kg^{-1} in rats administered orally.[13] (In comparison, the LD_{50} for aspirin, the common over-the-counter drug, is about 1500 mg kg^{-1} for rats). Atropine dilates eye pupils and is used in eye medicine. The species name "belladonna" relates to an historic use of deadly nightshade. Italian women used to squirt juice from the black berries into their eyes to obtain large pupils, an effect that was considered attractive.

Not all pharmacologically active plant compounds are alkaloids. The heart-active compounds from foxglove (*Digitalis* spp., Figure 5.21(a)), like digoxin **5.33**, are cardiac glycosides with steroid structures. Digitalis is widely used in heart medicine.

Many pharmacologically active plant compounds are unsuitable as medicines as they are too toxic or have unwanted side effects. Their original structures can be modified by chemical reactions into derivative compounds that have more desirable pharmaceutical properties. The story of the development of aspirin **6.16**, the common painkiller, is an illustration (Figure 6.10). Since ancient times, it had been known that chewing of willow bark (*Salix* spp.) could alleviate fever and pain, although this also caused an upset stomach. The fever-reducing compound in willows (and a few other plants) was found to be salicin **6.15**; it is a simple aromatic glycoside. Compare its structure with aspirin or acetylsalicylic acid **6.16**, and note the similarities (Figure 6.10). In 1897, straightforward chemical reactions were found that transformed the plant compound into aspirin, leading to a major product sold by the Bayer pharmaceutical company in Germany. Interestingly, aspirin does not only relieve fever and pain, but has been recognized to have a wide range of other medicinal applications in recent times.

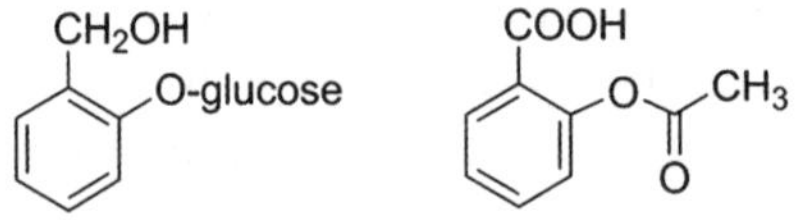

Figure 6.10 Salicin and aspirin.
Salicin **6.15**, a natural glycoside from willows, and aspirin (acetyl salicylic acid) **6.16**, the artificial pharmaceutical.

Plants and People 161

Among other uses, it has found applications as a blood-thinner and in the prevention of strokes and heart-attacks. Nowadays, aspirin is made completely synthetically. But the idea for its development came from a plant compound.

6.3.3 Newer Discoveries

In the 1950's, the American Cancer Institute initiated a widespread screening program of plant substances and extracts from various origins, as the potential of plants as sources for new pharmaceuticals was recognized. More than twenty years of research led to the discovery that an extract from the bark of the Pacific yew tree (*Taxus brevifolia*, Figure 6.11(a)) had anticancer activity.[14] The active compound was isolated. Paclitaxel, also known as taxol **6.17** (Figure 6.12), has been successfully used to treat ovarian, breast, and lung cancer. The compound undoubtedly has a complex structure. A big problem was that large numbers of Pacific yew trees were needed to produce only very small amounts of paclitaxel. Collection of the bark killed the trees, and repeated extensive use would have eventually led to the demise of Pacific yews. Other sources and methods were urgently needed. Fortunately, it was discovered that the needles of the much more common English yew (*Taxus baccata*) contain substantial amounts of a compound that can be further modified by chemical methods into paclitaxel.

(a)   (b)

Figure 6.11 Plant sources for new medicines.
(a) Trunk of Pacific yew (*Taxus brevifolia*) showing bark. (Photo by Tellur Fenner.) (b) Vine of wild yam (*Dioscorea* sp.) with tuber.

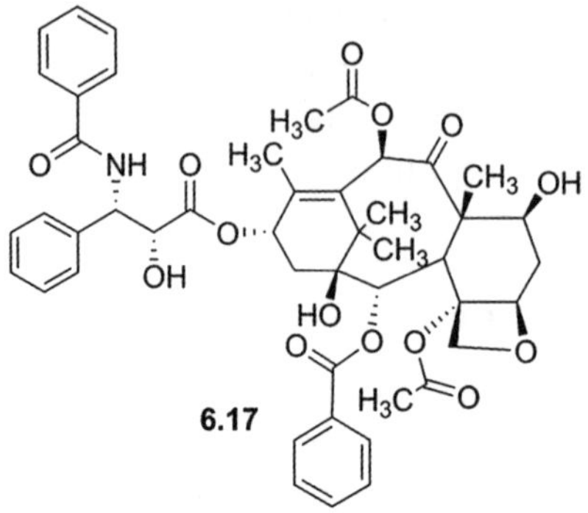

Figure 6.12 Paclitaxel, a recent discovery.
The complex structure of paclitaxel (taxol) **6.17**, an anticancer agent from Pacific yew trees (*Taxus brevifolia*).

The search for plant sources of steroids, in order to produce steroid hormones for medicinal applications, was an area of intense investigations in the middle of the last century. Earlier, steroidal products had to be obtained from animal sources, by slaughtering the respective animals and extracting the steroids, a tedious procedure. In search of suitable plant materials, American chemist Carl Djerassi and his coworkers discovered that tubers of wild yams (*Dioscorea* spp., Figure 6.11(b)) contained a saponin with a steroid structure, diosgenin **6.18**, that they could chemically transform into steroid hormones.[15] They were able to convert the plant saponin into cortisone **6.19**, a steroid used to treat inflammations (Figure 6.13). Diosgenin, the plant steroid, could be further used in the synthesis of progesterone **6.20**, a steroid hormone. It also provided starting materials for the synthesis of norethindrone **6.21**, the first oral contraceptive, which was accomplished in 1951. Note the structural similarities in the plant steroid, in the steroid hormones, and in the synthetic contraceptive, especially with respect to the ring connections in the steroid structures. (Check the wedged bonds pointing up at the chiral centers.) Synthesizing a complex pharmaceutical that has the correct steric arrangements is challenging. The plant steroid already provided the correct arrangements of the rings for an effective drug.

Plants have not only provided medicines through the ages. Their compounds also served as templates for the synthesis of new

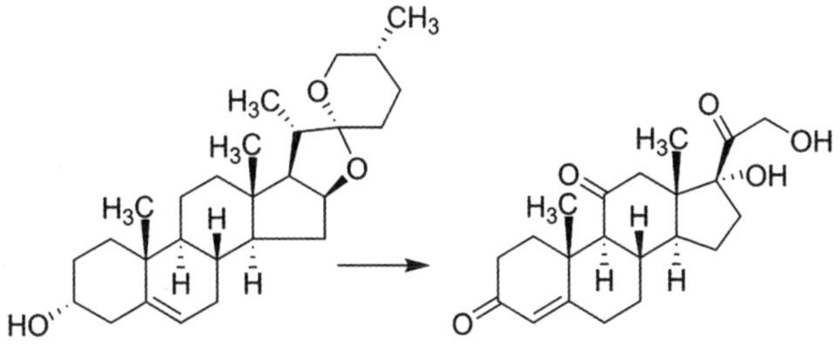

Plants and People 163

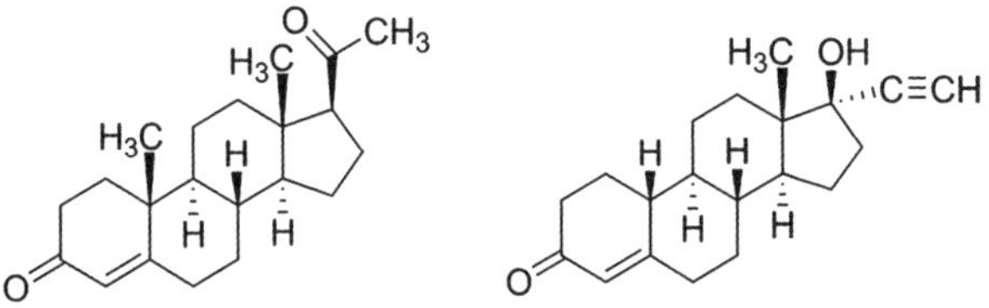

Figure 6.13 Plant steroids as sources for steroid drugs.
Diosgenin **6.18**, a saponin from wild yams (*Dioscorea* spp.). The human steroid hormones cortisone **6.19** and progesterone **6.20**. Norethindrone **6.21**, an oral contraceptive.

pharmaceuticals. Only an estimated 10% of the world's flowering plants have been thoroughly investigated for their chemical components and tested for pharmacological activities. A large number of plants are unexplored, especially plants from tropical rain forests. Conservation of this vast resource is of great importance. Resistant strains of human diseases keep developing, and new diseases appear. Plants may hold solutions for cures and provide inspiration for the synthesis of new effective pharmaceuticals.

6.4 PSYCHOACTIVE PLANTS

A small percentage of plants contain compounds that affect the mental state of humans, *i.e.* that are psychoactive (Figure 6.14).[16] Some of the plants and their compounds act as hallucinogens, meaning that they induce a perception or sensing of things that have no reality. Peyote and marijuana (cannabis) are examples. Other psychoactive plant compounds act as stimulants; they excite and enhance mental alertness and physical activity, reduce fatigue,

Figure 6.14 Plants with psychoactive compounds.
(a) Coca bushes (*Erythroxylum coca*). (Photo by Liselotte Wespe.)
(b) Joint fir (*Ephedra* sp.). (c) Hemp (*Cannabis sativa*).

or suppress hunger. Cocaine, ephedrine, and caffeine, are plant-derived stimulants with varying potencies. Yet other psychoactive compounds act as depressants that dull mental awareness, reduce physical performance, and induce sleep or a trance-like state. Alcohol and opiates like morphine are examples. Psychoactive drugs affect the central nervous system in various ways by influencing the release of *neurotransmitters*. The latter are natural chemical compounds that act as chemical messengers within the nervous system. Figure 6.16 will show some examples. After prolonged, heavy, regular use, many psychoactive compounds lead to addiction, which means they cause a physical and psychological dependence on the drugs. Discontinuing their use leads to severe withdrawal symptoms. Most psychoactive plant compounds are controlled substances.

Psychoactive plants have long been known to mankind. The use of hallucinogens extends back into prehistory. Many were used by shamans and medicine men in special ceremonies. A large number of psychoactive plants were also used as medicines, and many are still in use, in controlled doses, as part of modern pharmaceuticals. Chemical methods have been found that alter the chemical structures of the plant compounds and convert them into novel medicinal compounds. On the other hand, illicit chemical procedures can transform some of the plant compounds into much more potent psychoactive substances.

In plants, psychoactive compounds have defensive functions. As most of the compounds are alkaloids, their bitter taste repels animals. Some of the compounds are potent insecticides, like nicotine.

If an animal ingests compounds that alter perception, it cannot react properly to natural hazards and risks predation. Psychoactive plant compounds are primarily found in flowering plants (Angiospermae) and in some fungi. Some plant families are especially known for containing them: they include the nightshade family (Solanaceae) and the poppy family (Papaveraceae), plant families that we encountered earlier as common sources of alkaloids. A hallucinogen that is not an alkaloid is tetrahydrocannabinol (THC, **6.29**, see later in this section) from cannabis. Psychoactive compounds can occur in all parts of plants, in the fruits, the stems, the leaves, or the roots.

Some psychoactive alkaloids were shown earlier already, in a different context. Morphine **6.12** is a major component of opium, the latex obtained from unripe capsules of opium poppies (*Papaver somniferum*, Figure 6.8(a)). A synthetic derivative of morphine is heroin, a highly addictive substance that is not found in plants. Nicotine **5.36**, from tobacco plants (*Nicotiana* spp., Figure 5.24(b)), is a strong defensive compound in plants. In humans, nicotine acts both as a stimulant and as a relaxant. Regular use with inhalation leads to addiction. Pure nicotine is a highly toxic oily liquid. Its LD_{50} is 50 mg kg^{-1} for rats, administered orally. Caffeine **5.39**, from the coffee plant (*Coffea* sp., Figure 1.22(b)) is a bitter, white, crystalline compound that acts as a stimulant. Regular drinkers of strong coffee experience withdrawal effects when they stop drinking beverages that contain caffeine.

Figure 6.15 shows some additional examples of plant alkaloids that have psychoactive properties. The diversity of their structures is quite apparent. Yet, there are connections with earlier alkaloids and also with the structures of neurotransmitters. Cocaine **6.22**, isolated from the coca plant (*Erythroxylon coca*), is a potent stimulant and addictive substance. The coca plant has been playing a significant role in traditional Andean culture for centuries where chewing of coca leaves has helped people stave off fatigue and hunger. The practice combats altitude sickness, too. Recall that alkaloids are usually found as salts in plants. Native people have traditionally chewed coca leaves with small amounts of lime (burned limestone). This practice converts the cocaine salt into the basic cocaine alkaloid which is more easily absorbed through the mucous membranes, in a method of "freebasing". It was noted earlier that alkaloids are often classified according to their ring

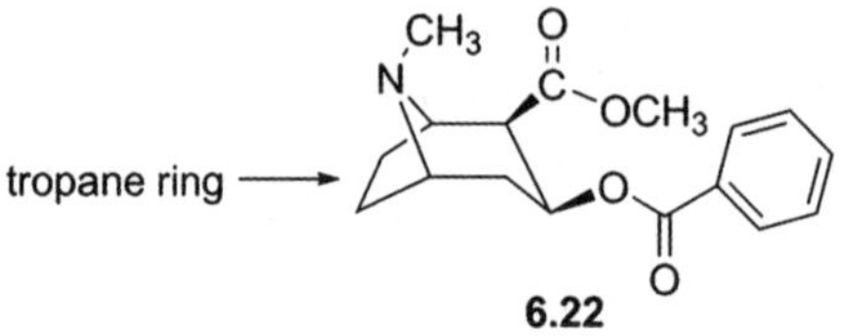

Figure 6.15 Psychoactive alkaloids.
Cocaine **6.22**, a tropane alkaloid, ephedrine **6.23**, mescaline **6.24**, and lysergic acid **6.25**, an indole alkaloid, are psychoactive alkaloids. (The indole ring system is shown in red.)

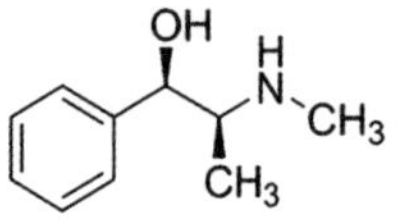

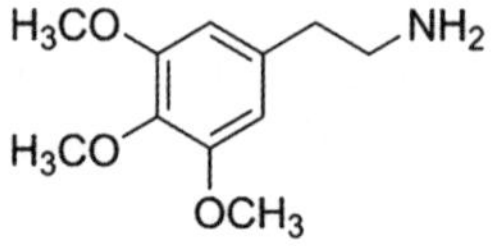

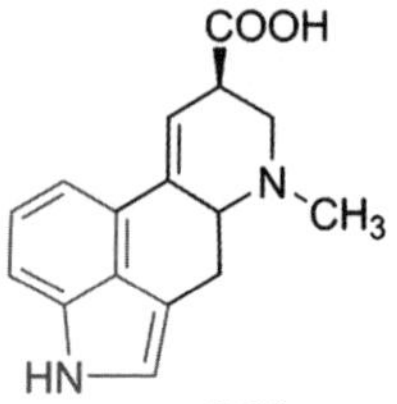

structures. Cocaine molecules have a large ring composed of seven carbon atoms and with a nitrogen group bridging it, called a tropane ring. It is indicated in Figure 6.15. This same ring system is also found in atropine **6.14** and scopolamine **5.37**.

The genus *Ephedra* (Figure 6.14(b)) is a type of shrub that grows in dry climates, like in southwestern North America, in southern Europe, in southwest and central Asia and in South America. The plants have long been known for their medicinal properties as stimulants, decongestants and appetite depressants. Ephedra is known as ma huang in China and is used in many Chinese herbal medicines. Different species of ephedra are known as joint-fir (referring to their segmented stems with only vestigial leaves) or as Mormon Tea. The latter name connects with the use of ephedra in the preparation of mildly stimulating teas by early settlers of the American Southwest. The psychoactive alkaloid in the plant is ephedrine **6.23**. As a dietary supplement for weight loss and enhancement of performance, ephedrine can have serious side effects and therefore is a controlled substance in the United States. There is a close relation of ephedrine to the structure of the neurotransmitter norepinephrine (noradrenaline) **6.26** (Figure 6.16). Synthetic amphetamines are psychostimulant drugs that are also used illegally under the street name "speed". They are structurally closely related to both ephedrine and norepinephrine.

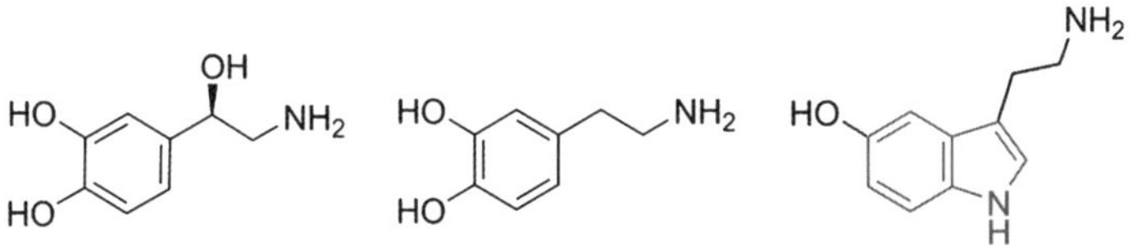

Plants and People

Figure 6.16 Structures of neurotransmitters.
Norepinephrine (noradrenalin) **6.26**, dopamine **6.27**, and serotonin **6.28** (with indole ring system highlighted in red).

Few cacti have toxic chemical defenses. (They usually have excellent physical defenses!). An exception is the peyote cactus (*Lophophora williamsii*) that grows in southern North America and parts of Mexico. Peyote has been used for over 3000 years by native people in religious ceremonies. The psychoactive compound in peyote is mescaline **6.24**, a bitter-tasting alkaloid with hallucinogenic activity. The compound has close similarities with the structure of the neurotransmitter dopamine **6.27**.

Rye plants (*Secale cereale*) can develop a fungal disease that causes black deformations of the grains. The disease is caused by the ergot fungus (*Claviceps purpurea*). This fungus produces toxic alkaloids that are various derivatives of lysergic acid **6.25**. During the Middle Ages, human poisonings due to the consumption of rye bread made from ergot-infected grain were common in Europe. The disease was known as St. Anthony's fire. Note the structural connections with the neurotransmitter serotonin **6.28**. Lysergic acid and serotonin have both an indole ring structure that consists of a benzene ring fused with a five-membered ring that has nitrogen in it. (The indole structure is highlighted in red in lysergic acid and in serotonin.) An artificial derivative of lysergic acid is LSD or lysergic acid diethyl amide. The compound was synthesized by Swiss chemist Albert Hofmann in 1938 while he was trying to find new drugs with analeptic (central nervous system stimulant) properties. To test the compound, he intentionally ingested a very small dose of LSD (a practice that was still in use then) and, as a frightening as well as fascinating experience, discovered that LSD was an exceptionally potent psychoactive producing strong hallucinations. LSD has been used in psychotherapy. The drug soon found its way into mainstream and became a part of hippie culture. LSD is an illegal drug.

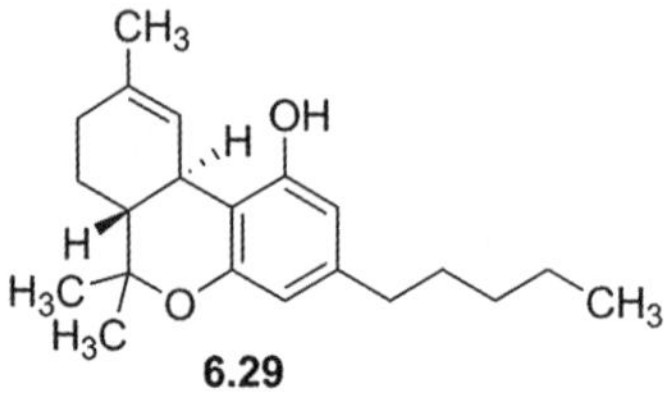

Figure 6.17 Structure of THC.
Δ⁹-Tetrahydrocannabinol (THC) **6.29** has a phenolic as well as an isoprenoid structure component.

The marihuana plant (*Cannabis sativa*, Figure 6.14(c)) has been used as a hallucinogenic plant since ancient times. Shamans made use of it in their ceremonies. The plant also had applications as a medicine. Nowadays, cannabis is the most widely used illicit drug. The typical herbal form of cannabis consists of the flowers, leaves, and stalks of female plants. Hashish is the concentrated resinous (and much more potent) form of the drug. Cannabis is very complex in its chemistry due to the large number of its constituents and their possible interaction with one another.[17] The best-known class of cannabis constituents consists of the cannabinoids, a class of (non-alkaloidal) compounds with twenty-one carbons and a phenolic and a terpene component. The psychologically most active compound of the mixture is Δ^9-tetrahydrocannabinol (THC) **6.29** (Figure 6.17). (Try to identify the isoprene units and the phenolic structure in THC.) The molecule has two chiral centers. THC is largely a hydrocarbon structure indicating that the compound is not soluble in water. The structure of THC was fully described in 1971. Even older than the use of cannabis as a hallucinogenic plant is its use as the source of hemp fibers, which leads us to the next section on plant fibers. While plants cultivated for drug use are bred for enriched THC content, hemp plants for the production of fibers are much lower in cannabinoids.

6.5 FIBERS AND DYES FROM PLANTS

6.5.1 Plant Fibers

Plants have been the sources of fibers for rope making, basket weaving, and clothing since prehistoric times. People used suitable, locally available plant materials to obtain fibers that could

Figure 6.18 Fiber plants.
(a) Cotton bolls (*Gossypium* spp.). (b) Stems and flowers of flax (*Linum usitatissimum*). (c) Desert yucca (*Yucca schidigera*) with fibrous leaves.

be twisted into ropes or woven into textiles.[18] Cotton plants (*Gossypium* spp.) have a wide distribution, and their ripe seed capsules, the cotton bolls (Figure 6.18(a)), have been a source of fibers for thousands of years. As early as 5000 BC, humans have used fibers from hemp plants (*Cannabis sativa*) to make ropes and to weave them into fabrics for making ship sails. Linen from the flax plant (*Linum usitatissimum*, Figure 6.18(b)) and ramie from an East Asian plant (*Boehmeria nivea*) have long been used to make textiles. Kapok, from the tropical kapok tree (*Ceiba pentandra*) has provided filling materials, like stuffing for pillows, whereas jute (from *Corchorus* spp.) and sisal (from *Agave sisalana*) have been used to make ropes and mats since ancient times. Yuccas (Figure 6.18(c)), bamboos, and palms, provided fibrous materials for making ropes and for basket weaving. All these plants continue to be important sources of fibers for many different uses.

Plant fibers can be derived from ripe seed capsules; examples are cotton bolls for cotton or seed fibers of the kapok tree for kapok. Other fibers come from plant stems, as from flax, hemp, jute, or ramie plants. In agaves or yuccas, leaves provide the source of fibers. Special preparations are needed to obtain fibers that are useful for spinning, weaving, or basket making. In cotton manufacturing, the seeds have to be separated from the fibers, followed by washing processes. Stems of flax plants need to be left rotting in water to remove the soft tissues, followed by washing, pounding and further mechanical processes, to obtain the linen fibers useful for textile making. Leaf fibers from yuccas need to go through pounding and stripping processes.

All fibers, whether natural or man-made, are composed of polymers. Long strands of polymeric macromolecules are connected by intermolecular bonds that shape the fibers and give them characteristic properties. The intermolecular links are commonly in the form of hydrogen bonds. Other intermolecular bonds, like sulfur cross-links or ionic interactions between the strands, are important in animal fibers. Plant fibers are mostly composed of cellulose and lignin. (For structures of cellulose and lignin refer to Chapter 2.3 and Figures 2.8 and 2.9.) Cotton fibers consist of more than ninety percent of cellulose. Waxes and proteins contribute to the outer layers of cotton fibers. Recall that cellulose is a polymeric carbohydrate or polysaccharide and has many hydrophilic OH groups. These functional groups make cotton a highly water-absorbent material. Natural fibers such as cotton can be chemically modified to form semisynthetic fibers, like rayon or acetate; they have somewhat different properties compared to the natural fibers.[19]

Plant fibers are also the sources for paper making. The technique was invented in China around 100 BC. To make paper, finely ground wood pulp (cellulose and some lignin) and other fibrous materials are suspended in water. The suspension is drained through a screen and then flattened and dried to produce paper. Hydrogen bonds link the short fiber pieces. The proportion of cellulose to lignin affects the quality and the type of paper obtained. As for other writing materials, papyrus plants (*Cyperus papyrus*) provided the sources for scrolls in ancient Egypt.

In contrast to plant fibers, animal fibers, like wool and silk, are composed of proteins. Unlike cellulose fibers, protein strands have acidic and basic sites in their molecules. Carboxylic acid groups in the proteins provide the acidic positions and amino groups the basic ones. (Compare with Chapter 2.5.2). Fully synthetic fibers, like polyester or nylon, are all string-like polymers with different monomers. Learning from the structures of natural fibers, scientists have been able to create artificial fibers, with properties useful for suitable applications. The different chemical structures of natural and artificial fibers have to be taken into account for their further treatment, be it for washing them or for applying dyes as shown next.

6.5.2 Plant Dyes, an Introduction

Humans have used plant materials to color their clothing since prehistoric times.[20–22] Earliest coloring applications were some mere stains from berries or roots, usually caused by the tannins in the plant materials. Later, plant materials, like bark, roots, or leaves, were steeped in water, and plant or animal fibers were immersed in the dye baths. With evolving civilizations, techniques that made use of plant extracts as dyes became more elaborate, and a large variety of colors could be obtained. Madder root, from the Mediterranean madder plant (*Rubia tinctorum*), dyed wool a bright red. The yellow stigmas of the saffron crocus (*Crocus sativus*) colored silk and wool a golden yellow. Deep-blue colors were obtained from treatment of fibers with extracts from the indigo plant (*Indigofera tinctoria*, Figure 6.19(a) and 6.19(b)) and, in a less pure form, from woad (*Isatis tinctoria*). Note that systematic names of classical dye plants often feature the species name "*tinctoria*" or "*tinctorum*" which means "of the dyers".

Dyes, whether from a plant or artificially made, are organic compounds that absorb light in the visible range and, in addition, have the ability to bond to a fiber or other material. Chapter 4 on plant pigments discussed the typical structural properties of molecules that absorb light. Like pigments, dyes have organic structures with long sequences of conjugated double bonds. Some functional groups, like C=O or OH-groups, enhance the absorption of light. In addition, dye molecules need to have

(a)   (b)   (c)

Figure 6.19 Sources of classic dyes.
(a) Indigo plant (*Indigofera tinctoria*), source of blue indigo. (b) Indigo dye bath. (c) Scale insects on prickly pear cactus are the source of cochineal, a red dye.

reactive sites that can form lasting bonds with fibers. Although flowers contain many colorful pigments, like the anthocyanins, most of their compounds bond poorly to fibers. Most lasting plant dyes come from leaves or roots. A useful dye must be colorfast, *i.e.* resist fading with washing or when exposed to light. Bonds between fibers and dyes may be in the form of ionic or covalent bonds, or they can be hydrogen bonds.

Many plant dyes color wool or silk well. Recall that animal fibers are proteins and have acidic and basic sites in their molecules. Dyes with basic properties can form ionic bonds with acidic sites in wool or silk, and vice versa, acidic dyes bond strongly with basic functional groups in protein fibers. Cotton, a polysaccharide, on the other hand, does not have acidic or basic functional groups and cannot form lasting bonds with most plant dyes. This can be remedied by pretreating cotton fibers with solutions of metal salts, so-called *mordants*. Metal ions, like iron, aluminum, or copper ions, easily bond with cotton fibers, forming *metal complexes* with the OH groups of cellulose. As a next step, the pretreated fibers are immersed in the dye solutions. The metal ions linked to cellulose then also bond to the dye molecules, forming strong links between fiber and dye. As an additional bonus, mordanting with different metal ions produces many different color shades. The technique can also be used with animal fibers. The mordanting of fibers has been used since historic times. The dye processes were carried out in copper or aluminum pots, or iron nails were added to the mixture, methods that were probably discovered accidentally.

Bonds between metal ions and organic molecules strongly absorb light and lead to an intensifying of color. We encountered this type of bond earlier in the natural molecules of the different types of chlorophylls.

6.5.3 Classic Plant Dyes

In historical times, the roots of the Mediterranean madder plant (*Rubia tinctorum*) were gathered for their red dye. It was the classic dye used to color the uniforms of soldiers of the British army, leading to the nickname "redcoats". The red dye is the compound alizarin **6.30**. It contains a highly conjugated anthraquinone structure, highlighted in red in Figure 6.20. Madder had to be used with mordants in order to provide a lasting color, and even then the

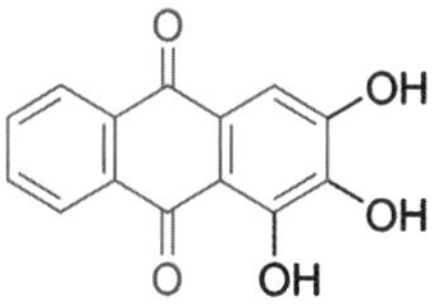

Plants and People 173

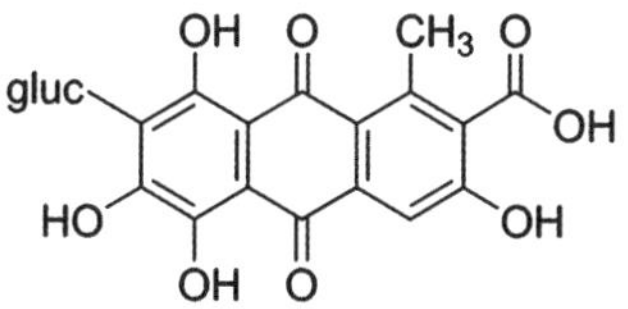

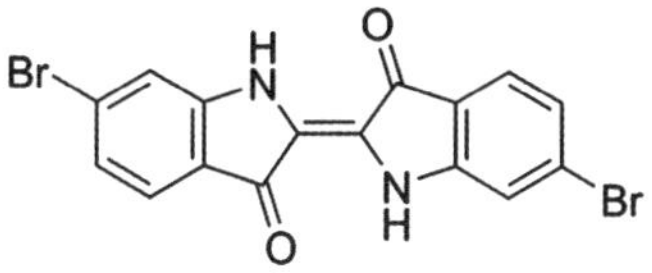

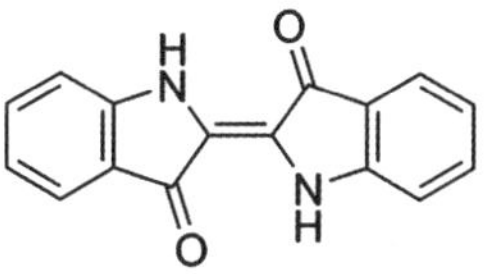

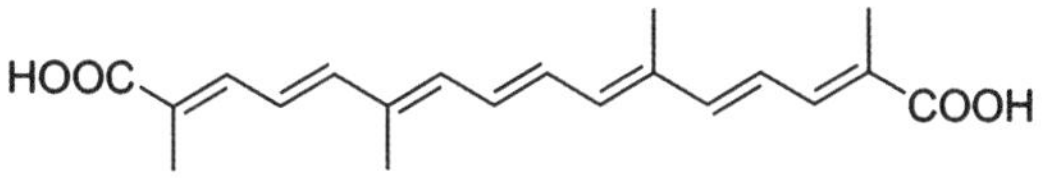

Figure 6.20 Structures of natural dyes.
Alizarin **6.30**, and carminic acid **6.31** both contain an anthraquinone structure. (It is highlighted in red in alizarin.) Tyrian purple **6.32**, and indigo **6.33**, two vat dyes, have indole structures. Crocetin **6.36**, $C_{20}H_{24}O_4$, is a terpene.

dye was not very colorfast. A large class of artificial dyes, the anthraquinone dyes, is related to alizarin. Synthetically made alizarin is still used in some inks and in staining techniques.

A better, but expensive classic red dye came from cochineal (*Dactylopius coccus*), a type of scale insect that lives on prickly pear cacti (*Opuntia* spp., Figure 6.19(c)) and that is native to Central America. The colorful compound is carminic acid **6.31**, with a chemical structure that is similar to alizarin. Note the carboxylic acid group in carminic acid which makes it bond well with protein fibers, although aluminum salts were commonly used as mordants for better color fastness. Cochineal is still used by native weavers of the Americas to dye wool for rugs. In the animals, the brilliant red dye acts as a feeding deterrent towards predatory insects.

It is interesting to note that animal and plant sources have produced some dyes that are very close in structure, as shown in the above example of carminic acid and alizarin. Similarly, Tyrian purple (also known as royal purple) **6.32**, was obtained from an animal source, namely the glands of marine snails. The process to obtain the dye was so costly that it was reserved for royalty, thus the name "royal purple". The chemical structure of Tyrian purple contains indole rings, introduced in Figure 6.15. The dye has a structure that is very similar to indigo **6.33**, the classic blue plant dye.

Indigo was obtained from extracts of the indigo plant (*Indigofera tinctoria*, Figure 6.19(a) and 6.19(b)). An elaborate process was necessary to transform the plant leaves into a dye that worked well on cotton. Indigo plants contain the colorless compound indican **6.34**, a glycoside (Figure 6.21). Fermentation of the leaves in strong base (classically in the form of ashes or stale urine) hydrolyzed the glycoside and formed the water-soluble, colorless leuco indigo **6.35**, "leuco" meaning colorless. Air oxidation of leuco indigo formed the deep-blue indigo dye **6.33**. Note the long conjugated system in the indigo molecule. Notice also that indigo is an alkaloid. Blue indigo is insoluble in water, but a dye needs to be dissolved in water in order to be absorbed by fibers. Cotton can absorb an aqueous dye solution very well. In order to apply the dye to the fibers, yarns and textiles had to be immersed in a solution of the leuco indigo and then exposed to air. Air-oxidation formed the water-insoluble indigo on the fiber. Alternatively, the mixture was (and in some places still is) transferred to large open containers, called "vats",

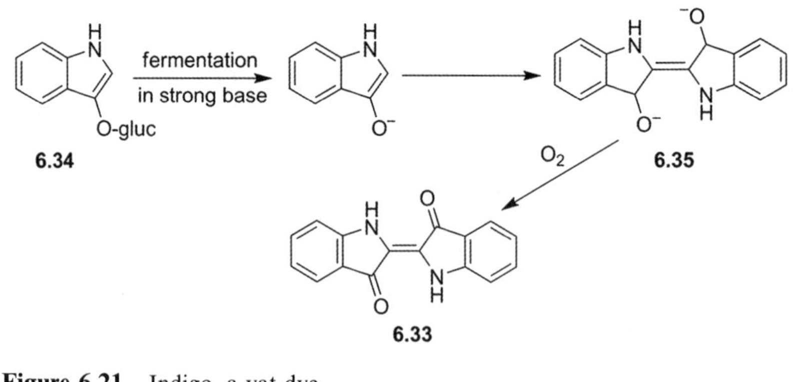

Figure 6.21 Indigo, a vat dye.
Indican **6.34** from the indigo plant, when fermented in strong base, forms colorless, water-soluble leuco indigo **6.35** that oxidizes in air into deep-blue—and water-insoluble—indigo **6.33**.

and stirred. Therefore, indigo is known as a "vat dye". During this process, air-oxidation set in, and water-insoluble indigo precipitated out. The settled dye was collected and pressed into cakes for trading. For dyeing fibers, indigo then first had to be transformed by a reduction process into the water-soluble leuco form to be absorbed by fibers, and then air-oxidized on the fiber. It can make us wonder how people figured out the multi-step procedure long ago. Indigo is still used, although mostly as a synthetic product, to dye millions of pairs of blue jeans.

Classic yellow dyes take us to terpenes. Crocetin **6.36** (Figure 6.20) is the deep-yellow dye in saffron, from the yellow stigmas of the saffron crocus (*Crocus sativus*). Thousands of flowers and many hours of laborious collecting them provide just a few grams of the stigma, which explains why saffron is the most expensive dye (and spice). Saffron colors silk and wool a lasting golden yellow. Note the two carboxylic acid groups. Saffron is still used to dye the saffron robes of monks. Saffron is a complex mixture of many different compounds. The typical aroma of the spice is due to smaller molecules, the monoterpene picrocrocin **6.8** and its aglycon safranal. The use of saffron goes back several thousand years, both for its dye and for its spice character, and its trade history is vast. The seeds from the shrubby plant *Bixa orellana* from tropical America contain another yellow isoprenoid dye. The compound's name is bixin, better known as "annatto", a natural food colorant that is used to color butter yellow in winter. As cattle feed in winter consists mostly of hay and not of the fresh greens that contain lots of carotenoids, cow's milk in winter and the butter produced from it are mostly white. But the public expects butter to have a certain color, therefore the addition of annatto. The seeds from *Bixa orellana* are also used as a spice and as a dye for fibers.

6.5.4 Plant Dyes and Contemporary Colorants

Until the middle of the 19th century, extracts from plants were the major sources of dyes to color textiles. In 1856, the English chemist Sir William Henry Perkin accidentally discovered the first artificial dye, as mentioned earlier, while trying to find a synthesis for the antimalarial quinine. He named the dye mauveine, for its purple or mauve color.[23] His discovery led to the beginning of synthetic dyes and brought largely an end to the use of plant sources for commercial dye processes. Very few plant colorants are still in use on a large scale. There are several reasons for this: most plant dyes color only wool and silk well and even these are not very colorfast, but good dyes must not pale with sunlight or washing. Due to season or geographic location, suitable plant materials may not be available. Also, the concentration of coloring compounds in plants varies. Therefore, repeated applications will not lead to even dyeing. Artificial dyes allow for an almost endless palette of colors, on many different materials, and can be weighed and applied to fibers for practically identical coloring.

Nevertheless, contemporary artists and home dyers enjoy experimenting with dye plants and appreciate the variability of the colors obtained on fibers. Materials from many locally growing plants, leaves, roots, or barks, are used for dyeing various types of fibers, and mordanting techniques add to the palette of colors obtained. Last but not least, plants provided the structural origins and inspirations for many of today's artificial dyes.

6.6 PERFUMES FOR PEOPLE

6.6.1 Introduction and History

Fragrant plant materials have contributed to perfumes, lotions, and oils since ancient times.[24-26] Scented woods, herbs, and spices had been used as incense in China, Egypt, Mesopotamia, and India thousands of years ago. People also learned how to extract fragrant substances from plant materials. In Egypt, flower petals were pressed with fats that drew the fragrant oils out of the flowers, and the mixtures were used as scented pomades. Lotions and perfumes were made by extracting leaves and flowers with alcohol. The fragrant materials were mostly reserved for nobility as they were expensive. Depictions of uses of aromatic plants have been found on hieroglyphs, and remnants of flower petals and scented woods were unearthed in ancient graves of Egypt and Mesopotamia. Frankincense and myrrh, gifts of the Magi, were important goods of trading. Through the crusades and through trading, the fragrant materials made their way west to Europe and became materials of luxury. A hue of luxury is still attached to good perfumes, shown by the carefully designed flasks for expensive fragrances.

The history of the isolation of fragrant plant compounds and the elucidation of their structures is similar to the developments of medicinal compounds from plant origins. Alchemists, like Paracelsus at the beginning of the 16th century, used distillation methods to isolate essential oils from plants that were known for their fragrances. Their explorations prepared the way for later isolations. Chemists of the 19th century were able to separate and identify numerous plant compounds, as described earlier in the section on plant medicines. These chemical investigations included many components of essential oils. As fragrant plant compounds are volatile and mostly water-insoluble, methods that allowed the isolation of plant fragrances included steam distillation and extraction with solvents like alcohol or hexane, a liquid hydrocarbon. Oils from the peels of citrus fruits, like lemons or oranges, could simply be squeezed out, in a method called "expression". Distillation, extraction, and expression are still used in perfume-making. An additional process, similar to the one used by the ancient Egyptians for creating pomades, is called "enfleurage" (from the French "fleur" for flower) in which fragrant flower petals are pressed onto purified fats. This makes the non-polar fragrant compounds diffuse into the fats. As enfleurage is a lengthy process, this technique is rarely used anymore.

Plants are the major sources for natural fragrant materials in the perfume industry. Flowers (like roses, jasmine, violets, and carnations), leaves (like mint and lavender, Figure 6.22(a)), seeds (like anise), and fruits (lemons and oranges) all provide ingredients for perfumes. So do some plant roots (*e.g.* ginger), barks (like cinnamon bark), and some woods (as from pines). Resins from two related African desert trees are the sources of the fragrant mixtures of myrrh and frankincense (Figure 6.22(b)). A few musky odors were historically obtained from animal sources, namely from the anal glands of the musk deer (*Moschus moschiferus*) from Central Asia and the civet cat (*Viverra civetta*). Nowadays, musky odors are produced synthetically.

Perfume ingredients from natural sources are expensive because plants contain only small amounts of the fragrant materials. Natural perfume components often spoil with time, either due to oxidation or moisture, or are degraded by acids on skin. Learning from the structures of natural materials, chemists of the 20th century developed numerous synthetic fragrant compounds.

Figure 6.22 Plants for fragrances.
(a) Lavender (*Lavandula angustifolia*). (b) The peeling bark of *Boswellia sacra* showing resins, the source of frankincense. (Photo by Sergej Maršnjak.)

The manmade scents greatly expanded the variety of fragrances and allowed the production of more stable, lower cost compounds, for use in inexpensive perfumes, or for articles like bath oils, deodorants, or shampoos. Synthetic ingredients in perfumes have long been frowned upon. This changed with the launching of the still best-selling perfume Chanel No. 5 in 1921 which, aside from natural oils such as from rose and jasmine, includes also synthetic fragrant compounds. The mixture of all the components together leads to the unique character of the perfume. Read on about some typical perfume ingredients.

6.6.2 Classic and Modern Perfume Ingredients

A large number of compounds that contribute to fragrances used in perfumery are, not astonishingly, terpenes and volatile aromatic compounds. In the chapters on attractive plant odors (Chapter 3) and on defensive scents (Chapter 5) we have already encountered many plant compounds that are part of natural perfume ingredients: geraniol **3.3**, limonene **3.4**, menthol **5.1**, α-pinene **5.3**, vanillin **3.5** and eugenol **3.6** are examples. A few additional perfume components from natural sources are shown in Figure 6.23. They serve as a reminder of terpene structures and compare two plant-derived odors with animal-derived scents. The monoterpene

Plants and People

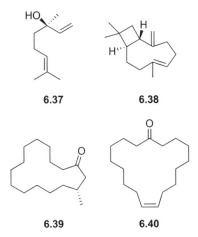

Figure 6.23 Natural perfume ingredients.
The terpenes linalool **6.37**, from lavender and rose oils, and caryophyllene **6.38**, from carnations and oil of cloves, are compounds with sweet fragrances. Muscone **6.39** and civetone **6.40**, originally from animal sources, provide musky notes.

linalool ($C_{10}H_{18}O$) **6.37** is found in rose and lavender oils. Caryophyllene ($C_{15}H_{24}$) **6.38** is a component of the fragrance of carnations and also of oil of cloves. Caryophyllene is a terpene with fifteen carbons. Terpenes consisting of three isoprene units and therefore fifteen carbons are generally known as *sesquiterpenes*. (Try to find the three isoprene units in caryophyllene.) Remember that volatile compounds consist of relatively small molecules and are mostly nonpolar. Fats as in enfleurage or solvents like hexane can therefore be used to extract the nonpolar fragrant compounds.

Compare the plant-derived terpene structures with muscone ($C_{16}H_{30}O$) **6.39** and civetone $C_{17}H_{30}O$ **6.40** (Figure 6.23), two animal-derived musky odor notes. Their structures do not contain isoprene units and therefore are not terpenes. The large hydrocarbon segments in the muscone and civetone molecules show that the compounds are fat-soluble.

As perfume components from natural sources are expensive, chemists in the perfume industry have developed methods to synthesize many of the compounds from sources like coal tar or wood. These processes are usually much less costly than the isolations from plant materials. Slight chemical alterations in the structures of compounds known from plant sources have led to new odor notes and sometimes to more stable compounds. Functional groups that are commonly found in synthetic (and natural) perfume compounds are aldehydes and esters. Aside from the systematic research to develop new fragrances by varying structures of known components with pleasing odors, some perfume ingredients were found by serendipity. In 1888, A. Baur accidentally discovered compounds with a musky odor while doing research on explosives related to TNT!

Perfumes are very complex compositions. While some ingredients evaporate quite fast on skin, other less volatile ones last a few hours. The least volatile components are the persisting odors (or what a garment smells like long afterwards). All the ingredients of a perfume must be dissolved in a solvent like alcohol, occasionally alcohol-water, in order to have the nonpolar essential oils in solution. Among other requirements, an acceptable perfume must have a lasting, pleasing color, has to be reasonably stable on skin, and must not be allergenic. Gas chromatography, often combined with mass spectrometry, is the technique of choice to determine the volatile components in perfumes. It is a most important technique in the quality control of fragrant mixtures.

As perfumes with synthetic fragrances can be made at a much lower cost than those obtained from natural sources, adulterated perfumes are sometimes sold fraudulently as natural ones, with corresponding high prices. The following methods to detect adulteration of perfumes illustrate some interesting comparisons between natural plant materials and artificial ingredients.

While natural fragrant oils contain hundreds of components, the composition of a perfume completely made from artificial compounds is a lot simpler, with much fewer components (and would show up a smaller number of peaks on a gas chromatogram). But, as a much more complex problem of fraud, perfume mixtures containing natural oils are sometimes diluted with cheaper synthetic fragrances. This can sometimes be detected by looking for by-products that can only be of synthetic origin. Another method takes into account that living things are in exchange with the natural atmosphere. Carbon dioxide in air naturally contains a very small, distinct percentage of radioactive carbon-14 (^{14}C), formed by cosmic radiation in the upper atmosphere. Plants incorporate this radioactivity through photosynthesis and through biosynthesis into further metabolites. On the other hand, coal tar, the source for making

Plants and People 181

many synthetic odorants, does not have this radioactivity as ^{14}C in coal tar has long decayed, and neither do synthetic fragrances produced from it. By determining the exact content of ^{14}C in a sample, unnatural sources of a perfume can be detected.

Research on perfume ingredients, whether natural plant materials or synthetic ones, has brought many new insights. It is a field of ongoing explorations and has led to the discovery of syntheses that are now applied in many other fields of chemistry. The creation of a successful perfume requires lengthy research and experimenting; it is an art!

6.7 GENETICALLY MODIFIED PLANTS

6.7.1 Introduction

Since the development of agriculture thousands of years ago, people have been sorting seeds of crop plants in order to grow plants with the most desirable traits. Seeds from plants that had produced ample crops or had otherwise outstanding properties were saved for sowing the following season. It is a custom that is still continued in many places of the world. For the last 200 years, learning from Gregor Mendel's experiments with plant hybrids and plant heredity, people have been selectively breeding and cross pollinating plants with desirable traits.[27] These processes led to recombinations within the genetic material, *i.e.* within the DNA in the cells. In general, years are required to obtain the desired results, and these breeding programs are successful only in related populations, *i.e.* in plants that can naturally interbreed.

Over the last few decades, as the result of research by thousands of scientists, the isolation of DNA from cells of virtually any organism has become feasible, together with the determination of the nucleotide sequences of their DNA. (Refer to Chapter 2.6 on Nucleic Acids for structures of DNA.) Furthermore, techniques have been developed to isolate genes, *i.e.* segments of DNA, that code for desirable traits from a host organism and to precisely insert them into another organism, even if host and receiving organisms are not related. The methods produce genetically modified organisms (GMO). Genetically modified plants, or GM plants, have new properties, such as freeze tolerance or a higher content of a vitamin.

Genetic engineering has brought major changes to agriculture.[28,29] With steadily growing human populations, increased crop yields are urgently needed. Genetic modifications have made it possible to grow crop plants in drought-stricken areas or on nutrient-poor soils. Some crops, such as maize, soybeans, potatoes, cotton, and sugar beets have been genetically modified to resist devastating insect pests. As a result, pesticide use can be reduced substantially. The technology has been applied to produce crop plants that are resistant to certain herbicides. It is also being used to increase the nutritional value of certain crops, such as rice, to alleviate nutritional deficiencies in people that have to rely on a limited variety of food plants.

6.7.2 Producing Genetically Modified Plants

In order to determine the location of desirable genes on the DNA of a plant, the make-up of the DNA, *i.e.* the sequence of its nucleotide base pairs, has to be determined first, in a method called DNA sequencing.[30] Recall that specific nucleotide bases are paired up in the double-stranded DNA, linked to each other by hydrogen bonding that pairs adenine (A) with thymine (T), and cytosine (C) with guanine (G). Particular nucleotide sequences are then determined for genes of interest, particularly for those that express desirable properties.

To obtain genetically modified plants, DNA sequences from a donor organism are selectively cut into smaller pieces with the help of *restriction enzymes* (Figure 6.24). These natural enzymes recognize specific nucleotide sequences in the double stranded DNA and cut the DNA wherever such a sequence occurs. The cuts in the sequence of nucleotides can be a few base pairs apart in the two strands, leaving so-called *sticky ends* (Figure 6.24(a)). With the help of another enzyme, called DNA ligase, these sticky ends can join with any other segment of DNA that has complementary base pairs, even from a different organism. DNA segments that are coding for desirable genes can thus be inserted into small DNA molecules, called plasmids, from a bacterium or virus: after restriction enzymes have been applied to cut both plasmid DNA and DNA containing a gene of interest, the DNA fragments with the sticky ends are spliced to form a *recombinant DNA* (Figure 6.24(b)). It has DNA that originated from different

Plants and People

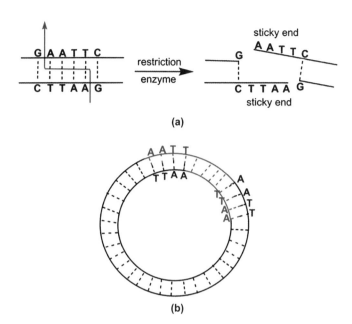

Figure 6.24 Restriction enzymes and recombinant DNA.
(a) A restriction enzyme cuts DNA after specific base sequences, leaving sticky ends. (b) With the interaction of the enzyme DNA ligase, a DNA fragment from a host plant is spliced with DNA from a bacterial plasmid by complementary base pairing, forming recombinant DNA.

sources, a donor and a bacterium (pointed out with different colors of the nucleotide bases in the figure). The new plasmid consisting of the recombinant DNA is picked up by the bacterial host cells and is rapidly replicated into millions of identical DNA, in a process called DNA cloning or gene cloning.

Several techniques are used to insert the modified DNA with the desired genes into the cells of a target plant. One of the techniques uses *Agrobacterium tumefaciens*, a species of bacteria that lives in the soil and that infects many flowering plants, including many crop plants. The bacterium reprograms plant cells by transporting its own genetic information into the attacked plant, causing plants to grow tumors. Therefore, the bacterium is an ideal organism to use for the transfer of desirable genetic information into plants. To produce GM plants, the genes known to induce the tumors are removed from the bacterial DNA and replaced with the desired genes. Plants rapidly integrate the modified genetic material.

To resist insect damage, major crop plants have been genetically engineered by inserting a gene from *Bacillus thuringiensis* (or Bt) into the plants' DNA. Bt is a soil-dwelling bacterium that can produce toxins that kill insects. Each cell of the GM plant expresses the insecticide, eliminating the need to spray crops with BT. Other crops have been genetically engineered to resist herbicides. GM crops that are thus engineered can be sprayed with herbicides and survive, whereas the competing weeds in the field are killed.

6.7.3 Impacts of Genetically Modified Plants

In 2007, farmers planted 114 million hectares of genetically modified crops worldwide. The United States are the top producers, growing nearly half of the GM plants, with the major crops being corn, soybeans, cotton, and canola.[31] But the environmental impact of GM crops continues to be a hotly debated topic, with some groups asking for stricter oversight on the use of GM plants and others for looser regulations. On one hand agricultural output must keep pace with steadily increasing populations, on the other it has to be accomplished without destroying ecosystems. Critics have objected to GM crops on several grounds, including ecological concerns, like the fear that pest-resistance is transferred to weeds, environmental impacts, like destroying habitat for biodiversity, and economic concerns raised by the fact that GM plants are subject to intellectual property law.

Close monitoring of the GM crops in use and their ecological impacts, together with a worldwide collection and exchange of data, are essential, in order to obtain well-documented information that allows well-founded decisions on the development of GM crops.[32-34]

6.8 CONCLUSION

Although the topics in this chapter describe widely differing uses of plants, there are common themes relating to the chemistry of plant compounds. The usefulness of a plant is closely tied to its chemical composition, the structures of its compounds, and the concentration of active compounds in the plant. Deciphering the chemical structures of plant compounds has brought much inspiration to research, be it to develop new medicines, or novel types of fibers, or a new component for a perfume. The change of just a few functional groups, be it in a medicine, drug, or dye, can profoundly change the properties of a compound, as in the example of morphine compared to codeine, in the fiber qualities of cotton compared to artificial rayon, or in the production of various colors in dyes. Slight synthetic changes from the structures of natural fragrances add new odor notes in perfumes.

The themes of this chapter applied many principles that were learned previously. Solubility properties are important in all uses of plant compounds and were specifically addressed in connection with vitamins, drugs, dyes, and perfume ingredients. Vitamins, herbs, and spices were shown to be potent antioxidants, compounds that can trap and inactivate highly reactive oxygen radicals that would otherwise destroy vital molecules. Small molecules of volatile compounds were encountered as part of food flavors and as essential components of perfumes, whereas polymers were shown as composing fibers. Long conjugated systems were met in the section on dyes. Major families of plant compounds, namely the alkaloids, contribute to plant medicines, drugs, and even dyes. Terpene structures compose some vitamins, drugs, and dyes; terpenes are major components of perfumes. Many of the structures presented in this chapter contain asymmetries in their molecules, and correct orientation of their chiral centers is essential for their proper activity. These compounds have been the focus of many challenging total syntheses. Finally, the topic of genetically modified plants built on a basic understanding of the structures of nucleic acids.

Plant compounds and their structures have been a major inspiration for research to find new useful materials for people. The quest to synthesize complex plant substances from simple compounds has led to many discoveries. Accidental discoveries have been part of the process: The first synthetic dye was discovered while attempting the synthesis of a medicine, and a new odor note for perfumes was found while working on explosives!

REFERENCES

1. H.-D. Belitz, W. Grosch and P. Schieberle, *Food Chemistry*, Springer-Verlag, Berlin, 3rd edn, 2004.
2. H. McGee, *On Food and Cooking: The Science and Lore of the Kitchen*, Scribner, New York, NY, 2004.

3. *Edible: An Illustrated Guide to the World's Food Plants*, The National Geographic Society, Washington, DC, 2008.
4. M. A. Asensi-Fabado and S. Munné-Bosch, *Trends Plant Sci.*, 2010, **15**, 582.
5. J. W. Finley, A. Kong, K. J. Hintze, E. H. Jeffery, L. L. Ji and X. G. Lei, *J. Agric. Food Chem.*, 2011, **59**, 6837.
6. A. O. Tucker and T. DeBaggio, *The Encyclopedia of Herbs*, Timber Press, Portland, OR, 2000.
7. M. J. Balick and P. A. Cox, *Plants, People, and Culture*, Scientific American Library, New York, NY, 1996.
8. S. J. Risch and C.-T. Ho (ed.), *Spices: Flavor Chemistry and Antioxidant Properties*, ACS Symposium Series, **660**, American Chemical Society, Washington, DC, 1997.
9. J. Sumner, *The Natural History of Medicinal Plants*, Timber Press, Portland, OR, 2000.
10. P. J. Houghton, *J. Chem. Educ.*, 2001, **78**, 175.
11. J. Mann, *Murder, Magic, and Medicine*, Oxford University Press, Oxford, 2000.
12. K. C. Nicolaou and T. Montagnon, *Molecules that Changed the World: a Brief History of the Art and Science of Synthesis and Its Impact on Society*, Wiley-VCH, Weinheim, 2008.
13. *Merck Index*, Merck & Co., Inc., Whitehouse Station, NJ, 14th edn, 2006.
14. D. G. I. Kingston, *Chem. Commun.*, 2001, 867.
15. C. Djerassi, *The Pill, Pigmy Chimps, and Degas' Horse*, Basic Books, New York, NY, 1992.
16. C. Raetsch, *The Encyclopedia of Psychoactive Plants*, AT Verlag Aarau/Switzerland, 2005.
17. M. A. ElSohly and D. Slade, *Life Sci.*, 2005, **78**, 539.
18. S. B. Warner, *Fiber Science*, Prentice-Hall, Englewood Cliffs, NJ, 1995.
19. T. M. Letcher and N. S. Lutseke, *J. Chem. Educ.*, 1990, **67**, 361.
20. M. Séquin-Frey, *J. Chem. Educ.*, 1981, **58**, 301.
21. R. M. Christie, *Colour Chemistry*, The Royal Society of Chemistry, Cambridge, 2001.
22. H. Zollinger, *Color Chemistry*, Verlag Helvetica Chimica Acta, Zürich, and Wiley-VCH, Weinheim, 2003.
23. S. Garfield, *Mauve: How One Man Invented a Color That Changed the World*, W. W. Norton & Company, New York, NY, 2001.

24. C. Sell, ed., *The Chemistry of Fragrances: From Perfumer to Consumer*, The Royal Society of Chemistry, Cambridge, 2006.
25. A.-D. Fortineau, *J. Chem. Educ.*, 2004, **81**, 45.
26. P. Loyson, *J. Chem. Educ.*, 2011, **88**, 146.
27. R. M. Henig, *The Monk in the Garden: The Lost and Found Genius of Gregor Mendel, the Father of Genetics*, Houghton Mifflin, Boston, 2000.
28. J. A. Thomson, *Seeds for the Future: The Impact of Genetically Modified Crops on the Environment*, Cornell University Press, Ithaca, NY, 2007.
29. S. O. Duke, *J. Agric. Food Chem.*, 2011, **59**, 5793.
30. M. Y. Herring, *Genetic Engineering*, Greenwood Press, Westport, CT, 2006.
31. E. Youngstedt, E. Stokstad, K. Krause and P. Huey, *Science*, 2008, **320**, 466.
32. M. Marvier, Y. Carriere, N. Ellstrand, P. Gepts, P. Kareiva, E. Rosi-Marshall, B. E. Tabashnik and L. LaReese Wolfenberger, *Science*, 2008, **320**, 452.
33. J. Davidson, *PlantSci.*, 2010, **178**, 94.
34. B. Pöpping, *J. Chem. Educ.*, 2001, **78**, 752.

Epilogue

This book has attempted to develop an understanding of chemistry, particularly of organic chemistry, by closely examining plant substances and their uses. The great diversity of plant compounds and their connections with our daily lives elicit admiration and inspiration. Not only do our lives depend on plants and on the chemical substances they make, but plant compounds have provided crucial ideas for new medicines, fibers, foods, and fragrances. The history and evolution of chemistry is closely tied to research on plants and their compounds. It is hoped that this book has created a deeper understanding of plants and their chemical components and an appreciation that extends to everyday observations of the natural world.

Glossary

How to Understand Plant Names

The binomial system of plant names (based on Carl Linnaeus' work) assigns a specific name to each species of plant and consists of genus (*pl.* genera) and species.

Example: *Rosa rugosa* (a type of rose). Genus: *Rosa*, species: *rugosa*. Another species: *Rosa californica*.

Abbreviation of species: sp. (*pl.* spp.). Sp. is sometimes used when a species is not defined.

Genera are grouped into families. Example: the Rose family (Rosaceae) includes numerous other plants aside from roses, all with certain common characteristics.

Plant families are grouped into orders.

Glossary of Terms

Accessory pigments: Light-absorbing pigments that work in conjunction with chlorophyll.

Aerobic respiration: Respiration in the presence of plenty of oxygen, leading to the complete breakdown of sugars and other organic compounds into carbon dioxide and water.

Aglycon: The non-sugar portion of a glycoside.

Alkaloids: A large family of nitrogen-containing organic natural products with basic properties.

Anaerobic respiration: Respiration in the absence of oxygen.

Anions: Negatively charged ions.
Anthocyanins: A large group of water-soluble phenolic plant pigments. A subgroup of flavonoids.
Antinutrients: Compounds that interfere with the absorption of nutrients.
Antioxidants: Molecules that inhibit the oxidation of other molecules by trapping radicals.
Aqueous: Dissolved in water.
Atomic number: The unique number of protons in the nucleus of each atom of an element (equal to the number of electrons in the neutral atom). Determines the position of an element in the periodic table.
Atoms: The smallest particle of an element that retains the chemical nature of the element.
Bases: Substances that produce an increase in the concentration of OH⁻ ions when dissolved in water. Opposite of acids.
Basic or **alkaline:** Producing hydroxide ions (OH⁻) when dissolved in water. Having a pH above 7.
Betalains: A family of nitrogen-containing plant pigments that are found only in specific plant families.
Biosynthesis: The biological synthesis of natural products.
C_3 plants: Plants that follow only the Calvin cycle, or C_3 pathway, in the fixation of CO_2; the first stable compound is a three-carbon compound. It is the most common pathway of CO_2 fixation.
C_4 plants: Plants in which the first product of CO_2 fixation is a four-carbon compound. It is a pathway common in plants originating from hot climates.
Calvin cycle: The series of light-independent photosynthetic reactions during which CO_2 from air is reduced to produce organic molecules in the form of simple sugars.
CAM plants (crassulacean acid metabolism): A variant of the C_4 pathway, characteristic of most succulent plants, such as cacti.
Carbon skeleton: The carbon-carbon backbone of an organic molecule.
Cardenolides or **cardiac glycosides:** A group of steroid glycosides from plants that affect the activity of the heart muscle.
Carotenoids: A group of C_{40} isoprenoid pigments.
Catalyst: A substance that increases the rate of a reaction without being used up in the process.
Cations: Positively charged ions.

Chiral: Asymmetric or "handed".

***Cis/trans* isomers:** Stereoisomers that differ in the positioning of atoms attached to a carbon-carbon double bond (or a ring).

Coagulate: To change to a solid or semisolid state, to form clots.

Cofactor: A nonprotein component that is required by enzymes in order to function.

Compound: A substance composed of two or more elements that are chemically combined in fixed proportions.

Condensed tannins: Large, complex phenolic molecules, providing dark pigments.

Conjugated double bonds: Alternating single bond-double bond sequences.

Coordinate covalent bond: A covalent bond formed when one atom donates both electrons for a bond. Common in metal-organic bonds.

Covalent bonds: Bonds formed by sharing of electrons between two atoms.

Cuticle: Waxy or fatty layer on outer wall of epidermal cells of leaves and stems.

Cyanogenic glycosides: A group of plant glycosides that release HCN when reacting with enzymes from neighboring plant cells.

Denaturation: Destruction of the original shape of a protein which makes it non-functioning.

Disaccharides: Carbohydrates consisting of two sugar units. An example is sucrose.

Dissociation: Separation into ions.

Diterpenes: Terpenes consisting of four isoprene units and thus having twenty carbon atoms.

Dye: A colorant, like a plant pigment, that can form bonds with a fiber like wool or cotton.

Electron configuration: The arrangement of electrons in defined energy levels around an atom's nucleus.

Electronegative: Having a tendency to attract electrons.

Electrons: Negatively charged particles, arranged in specific energy levels, occupying the space around an atom's nucleus.

Element: A substance composed of only one kind of atoms.

Enantiomers: Compounds that are mirror images of each other.

Endergonic: Describing a chemical reaction that requires energy to proceed.

Endothermic: Describing a chemical reaction that requires energy in the form of heat to proceed.
Enzyme: A biological catalyst.
Essential amino acid: An amino acid that is not produced by an organism and that has to be ingested through nutrition.
Essential oils: Volatile, fat-soluble plant oils (from the word "essence").
Exergonic: Energy-yielding (as in chemical reactions).
Exothermic: Giving off energy in the form of heat.
Exudate: A fluid or sap that is formed as a response to injury.
Fermentation: The extraction of energy from organic compounds in the absence of oxygen.
Flavonoids: A large family of plant pigments with a common phenolic three-ring structure.
Functional groups: Attachments to the carbon skeleton that provide characteristic chemical properties to compounds.
Glucosinolates: A family of sulfur-containing, defensive plant glycosides.
Glycolysis: The breakdown of sugars during respiration.
Glycoside: A non-carbohydrate molecule that is bonded to a sugar.
Gum: Water-soluble plant exudates that consist of large carbohydrate structures, usually formed as a response to injury.
Herbivores: Organisms that feed on plants.
Hydrated: Surrounded by water molecules. Abbreviated as (*aq*).
Hydrocarbons: Organic compounds consisting of hydrogen and carbon.
Hydrogen bonds: A chemical bond formed through the attraction between the partial positive charge on a hydrogen atom and the partial negative charge on a nearby oxygen or nitrogen atom.
Hydrolysis: Cleaving a molecule by reaction with water.
Hydrolyzable tannins: Water-soluble, phenolic plant compounds derived from gallic acid.
Hydrophilic: Having an affinity for water. (Opposite of hydrophobic.)
Hydrophobic: Water-repelling, fat-soluble.
Inorganic compounds: Compounds that do not contain carbon, as well as carbonic acid, carbon dioxide, carbon monoxide, carbonates, cyanides and a few others.
Ion: A charged atom.
Ion-exchange: Exchange of one type of ion with another one.

Ionic bonds: Bonds between positively and negatively charged ions, formed due to electrostatic attraction.
Ionic compounds or **salts:** Compounds with ionic bonds.
Isomers: Molecules with the same molecular formula but with different structures.
Isoprene unit: Characteristic C_5 building block of terpenes.
Isoprenoids: Another name for terpenes.
Isotopes: Atoms of the same element, but with different numbers of neutrons.
Latex: A milky plant sap that contains rubber particles.
LD_{50}: Lethal dose that kills 50% of a population of test animals, like mice or rats; a measure of toxicity of a compound.
Light reactions: The reactions of photosynthesis that require light.
Lipid: Any non-polar substance, found in biological systems, that is fat-soluble, but insoluble in water.
Macronutrients: Inorganic chemical elements required in large amounts for plant growth, like nitrogen, potassium, phosphorus.
Metabolism: The sum of all chemical reactions occurring in living organisms.
Metabolite: Compounds formed by living systems.
Metals: Elements that conduct heat and electricity well. Tend to form positive ions. Metals are found on the left hand side of the solid dividing line in the periodic table.
Micronutrients: Essential inorganic chemical elements required only in trace amounts, such as iron, manganese, zinc.
Mineral: An inorganic substance other than water. Usually an inorganic salt or its ions.
Molecular formula: A formula that shows the total number of atoms of each element in a molecule. Example: $C_6H_{12}O_6$.
Molecule: A particle consisting of two or more atoms joined by a chemical bond.
Monomers: The subunits that serve as building blocks in polymers.
Monosaccharides: The simplest carbohydrates or sugars, consisting of one sugar unit. Example: glucose.
Monoterpenes: Terpenes consisting of two isoprene units and thus having ten carbon atoms.
Mordant: A solution, usually of a metal salt, used to pretreat fibers before the dye process. Metal ions bond to the fiber and then also to the dye molecules.

Neurotransmitters: Natural chemical compounds that act as chemical messengers within the nervous system.
Neutrons: Neutral particles in the nucleus of an atom.
Non-metal element: Element that lacks metallic properties. Found on the right hand side of the solid dividing line in the periodic table. Tends to form negative ions.
Nucleotides: The monomers of nucleic acids (DNA or RNA).
Orbitals: The three-dimensional spaces where electrons can be found about 90% of the time.
Organelle: Specialized subunit within a plant cell with a specific function. Examples: cell nucleus, chloroplast.
Organic compounds: Compounds that contain carbon in their structure. (For exceptions see "inorganic compounds".)
Organo-metallic compounds: Compounds that have bonds between a metal ion and an organic portion of the molecule.
Osmosis: The diffusion of water across a selectively permeable membrane.
Oxidation: Reaction in which a compound or ion loses electrons. An oxidation has taken place if a compound has gained oxygen atoms or has lost hydrogen atoms.
Oxygenic (as in oxygenic photosynthesis)**:** Generating oxygen.
Peptide: A molecule that is made up of two to a couple of hundred amino acid monomers. Not as large as proteins.
pH: A measure that indicates how large the surplus of H^+- or OH^--ions is in an aqueous solution. The pH-value is defined as the negative logarithm of the hydrogen ion concentration. The pH range extends from 0 to 14, with 7 being neutral.
Phenolics: Organic molecules that have phenolic rings, *i.e.* hydroxyl groups (OH) attached to an aromatic ring.
Phloem: The part of vascular tissues that transports sugars and other organic compounds in a plant.
Phospholipid: Phosphorylated lipid. Similar in structure to a fat, but with only two fatty acids attached to the glycerol backbone, with the third space occupied by a phosphorus-containing molecule. Phospholipids are important components of cellular membranes.
Phosphorylation: A reaction in which phosphate is added to a compound.
Pigment: A compound that absorbs a section of sunlight and traps this way energy for further reactions.

Polar compound: Compound consisting of molecules with an unequal distribution of charge.
Polymers: Giant molecules that consist of a large number of repeating subunits (monomers).
Polysaccharides: Polymeric carbohydrates. Examples: starch, cellulose.
Primary metabolites: Molecules that are found in all plant cells and that are necessary for plant life.
Products: Substances formed as the result of a chemical reaction.
Proteins: Polymers that consist of amino acid monomers linked up by peptide bonds.
Protons: Subatomic particles in the atomic nucleus, each with one positive charge.
Racemic mixture: A mixture of equal amounts of two enantiomers.
Radicals: Atoms or molecules that have unpaired valence shell electrons.
Radioactivity: Emissions resulting from the spontaneous decay of unstable atomic nuclei.
Reactants: Starting substances in a chemical reaction.
Reduction: Reaction in which an ion or molecule gains electrons. This may also involve the loss of oxygen atoms and/or the gain of hydrogen atoms.
Resin: A water-insoluble, elastic plant exudate.
Respiration: The intracellular process in which larger organic molecules are broken down to smaller ones, usually in the presence of oxygen, with the help of enzymes.
Salt: Generally the same as ionic compound. Specifically often refers to sodium chloride (NaCl).
Saponins: A group of soapy, surface-active defensive plant compounds.
Secondary metabolites: Metabolites derived from the primary metabolites, restricted in their distribution in plants and often specific to certain types of plants.
Serpentine (serpentinite): Rocks and soils with low content of macronutrients like calcium, and high in magnesium content and heavy metals (a property referred to as "ultramafic").
Sesquiterpenes: Terpenes consisting of three isoprene units, thus having fifteen carbon atoms.

Stereoisomers: Molecules with the same connections of atoms but with different orientation of groups in space. Examples: Enantiomers, *cis/trans* isomers.
Steroids: A large family of organic natural products with a characteristic four-ring pattern.
Stigma: Part of a flower that receives pollen from pollinators.
Stoma, *pl.* **stomata:** Small opening in the epidermis of leaves and stems through which gases can pass.
Strong acid/strong base: Acid or base that is fully dissociated into ions in water.
Tannins: Acidic, phenolic plant pigments with astringent properties.
Terpenes: Secondary metabolites composed of isoprene units.
Total synthesis: Synthesis of a complex compound from simple, easily available organic compounds.
Transition elements: Metallic elements listed in the short vertical groups in the center of the periodic table.
Unsaturated: Having double (or triple) bonds.
Vacuole: A space or cavity bounded by membranes within the cytoplasm of a cell, filled with a watery fluid.
Valence electrons: Electrons in the outermost electron shells of an atom.
Vascular system: The system of conducting tissues in plants, including the xylem and the phloem.
Volatile: Evaporating easily.
Weak acid/weak base: An acid or base that is only partially dissociated into ions in water.
Xanthophyll: A yellow plant pigment related to carotenoids.
Xylem: Vascular tissue through which most of the water and minerals of a plant are conducted.

Photo Credits

All photos are by Margareta Séquin, unless credited otherwise in the figure captions.

Cover photo: *Rhododendron arboreum* (University of California Botanical Garden, Berkeley). Photo by M. Séquin.

Subject Index

Note: The index covers the main text but not the Glossary. Page references in *italics* indicate that the subject is only mentioned on that page in a Figure or a Table.

abietic acid 120
accessory pigments 29, 31, 34, 96, 98
Acer macrophyllum 59
acetyl choline 131
acetyl CoA 74–5, *76*
acidity
   defensive tastes and 123–6
   pigment color and 105
   soils 24–5, *26*, 109
acids and bases 21
addiction 164–5
adenine 71, *72*
aerobic respiration 37
Aesculus californica 111, *135*, *137*
Agave spp. 33, 135–6
   A. shawii 115
   A. sisalana 169
aglycons 128, *136*, 175
Agrobacterium tumefaciens 183
alanine 67, *68*
alcoholic fermentation 39
   see also ethanol
aldehydes
   benzaldehyde 134
   in chlorophyll 96

   in flavors 154–5
   in odors and perfumes 86, 91, 180
algae 98
alizarin 172–3
alkaline soils 25
alkalis *see* bases
alkaloids
   betalains as 105, 107
   caffeine as 52–3, 140
   indigo as 174
   medicinal *158*, 159
   morphine as 140
   nicotine as 140
   as plant defenses 140–2
   psychoactive 165, *166*, 167
   quinine as 128, 140
   taxine 115
allelopaths 132–3
Allenrolfea occidentalis 26
alliin and alicin 122, *123*
Allium spp. 122–3
   A. falcifolium 7
allyl isothiocyanate 121, *123*
almond trees 134

Subject Index 199

aluminum 105
Amanita muscaria *106*, 107
amber *119*, 121
amines 86
amino acids 65–8, 72, 148
Amorphophallus titanum 37–8, 79, 86
amphetamines 166
amygdalin 134
amyl acetate 86
anaerobic respiration 38
analytical techniques
  odors 88–90
  perfumes 180–1
animals
  alkaloids in 141
  carotenoids in 99
  fibers from 170, 172
  perfumes from 177
  seeds spread by 85, 88, 116
anions 7, 13
annatto 175
anthocyanins 102–5
  betalains and 107
  in fall coloration 111
anthraquinones 172–3
antinutrients 127
antioxidants 101, 111, 121, 150–3, 185
apples 85, *86*, 124, 133
Arachis hypogea 60
Arctostaphylos sp. 53
Argemone sp. 100
aroma *see* odors
aromatic compounds 57
  odors in 84–5
  in perfumery 178
  in spices 155
Arum family 37–8, 125–6

Asclepias spp. 128, 130, 137–8
  A. speciosa *138*
  A. tuberosa *139*
ascorbic acid *149*, 150
aspartic acid *68*
aspirin 160
Astragalus praelongus 6, *7*
asymmetric molecules 51–2, 67, 69, 118, 150
  see also chirality
atomic structure 9–13
  chemical bonding and 13–15
ATP (adenosine triphosphate) 31, 37–8, 73–4
  ADP and AMP 73–4
Atropa belladonna 3, 140, *158*, 159
atropine 140, *158*, 159
autumn *see* fall coloration

bachelor buttons 102, *103*
Bacillus thuringiensis (Bt) 184
bacteria in genetic engineering 183
'ball-and-stick' models 14, *16*, 81–2
bananas 85, *86*
barrel cactus 32
bases 21
  alkaloids as 22, 141
bat pollination 53, 108
bees 53, 101, 107–8
beetroots 105–6
benzaldehyde 134
Bermuda grass 32
Beta vulgaris 105–6
betacyanins 106
betalains 105–7
betaxanthins 106
bibliography 45–6

bigleaf maple *59*
biomembranes and lipids 65
bird pollinators 53, 107
bitter tastes 126–8
bitterroot 126
Bixa orellana 175
black pepper 152, 154
black walnut trees 132
blue-gum eucalyptus *94, 117*
Boehmeria nivea 169
boron 7
Boswellia sacra 178
bougainvillea 107
Brassica juncea 122
Brassicaceae 121–2
bristlecone pine *109, 119*
brown rice 149
buckeye tree *111, 135,* 137
bull kelp *29*
bush pickleweed *26*
butterflies 53, 95, 108, 122, 139

C_3 and C_4 pathways 32
cabbage family 121–2
'cabbage white' butterflies 122
cacti
    adaptation to drought *32,* 33, 64
    defense mechanisms 114, 167
    pigments *94, 106,* 107
    scale insects 173
cadaverine *40,* 41, 86, *87*
caffeine 52–3, 140, *141,* 142, 164–5
calcium
    ions 13, 24, 124–5
    as a macronutrient 4, 6–7
calcium carbonate 22, 96
calcium oxalate *124,* 125
California buckeye *135*

California poppies 97
Calla lily 125
calotropin 138
Calvin Cycle (Melvin Calvin) 30–3
CAM (crassulacean acid metabolism) plants 32, 33
cancer treatments 161
cannabis 163, 165
Cannabis sativa 168–9
canola 184
Cape sorrel *100*
capillary action 17
capsaicin *154,* 155
capsanthin *154*
Capsicum annuum 153, 155
carbohydrates 49–56, 146–7
    see also sugars
    cellulose 56–9, 170
    starch 54–5, 147
carbon
    atomic structure 9–10, *12*
    chemistry of compounds (*see* organic chemistry)
    fixation 30–3
    line structure representations 33–4, 41
carbon-14 180–1
carbon dioxide
    in photosynthesis 7
    in rainwater 21–2
carbon skeletons 40–1
    fatty acids 74
    terpenes 82–4
carbonates 22
carcinogens *131,* 132
cardiac glycosides (cardenolides) 51, 130, 136–9, 160
carminic acid 173
carnations 84, *85,* 179

β-carotene 97–9, *149*, 151
carotenoids 34, *35*, 84, 97–100
   see also terpenes
   in animals 99
   xanthophylls *98*, 99–100
carrots 97, *149*, 152
Caryophyllales 107
caryophyllene 179
cassava roots 134
castor bean plant 66
catabolism 37
Catalina cherry *133*, 134
catalysis 28, 190
   see also enzymes
cations 13, 23–4
Ceiba pentandra 169
cellulose 56–9, 170
Centaurea cyanus 102, *103*
chemical bonding
   see also covalent bonds; hydrogen bonds
   molecule formation 13–15
   in organic chemistry 40
   peptide bonds 68–9
chemical reactions
   in plants 27–8
   reaction pathways 74–6
Chenopodiaceae 107
chili peppers 153, 155
chirality 118
   see also asymmetric molecules; stereoisomers
   in alkloids 141, 159
   in amino acids 67, *68*
   in ascorbic acid 150
   in carbohydrates *51*, 52, 54
   and geographic location 118
   and odor 83
   pharmaceuticals 157, 159, 162, 185
   in steroids 136, 138, 162
   in terpenes 83, 99, 132
   THC 168
chlorine 7
Chlorogalum pomeridianum 135
chlorophyll 1
   chlorophyll *a* 33–4, *35*, 96–7
   chlorophyll *b* 34, *35*, 96
   chlorophyll *f* 96
   isoprene units 84
   in photosynthesis 29–30
   as a pigment 95–7
   plants without 36
chloroplasts 95, 97
cholesterol 136, 148
chromoplasts 95, 99
chrysanthemum 100
Cinchona tree 126, 128
   C. pubescens *158*, 159
1,8-cineole 117, *118*
cinnamaldehyde 154–5
Cinnamomum aromaticum 152
cinnamon 152, *153*
cis and *trans* fats 63, 147–8
citric acid 124
Citrus spp. *44*, 81–2, 124, 177
   vitamin C from *149*, 150
civet cat 177
civetone 179
Claviceps purpurea 167
cloves 84, 152–4, 157, 179
Cnidoscolus angustidens 130–1
coast live oak *109*
Coast redwood trees 56, *126*, 127
Cobra Lily 5
coca plant 126, 165
cocaine 164–6
cochineal *171*, 173
codeine 158–9
Codiaeum variegatum 129

coffee plants (*Coffea* sp.)
   23, 140, *141*, 165
Colocasia esculenta 125
colophony 120
color
   see also pigments
   dyestuffs 170–6
   of insects containing
      toxins 139
   of malodorous plants 86
competition and growth
   repression 132–3
compost 24
   see also soils
coniferous trees 108–9, 119
coniferyl alcohol 57, *58*
coniine 140–1
Conium maculatum 140
conjugated double bonds
   in aromatic compounds 57
   in chlorophylls 34, 94–5
   in dyestuffs 171–4
   in juglone 132
   in other pigments *98*, 99,
      103, 105–6, 155
coordinate covalent
   bonds 34
copper 4, 7, 172
coral root *36*
Corallorhiza sp. *36*
Corchorus spp. 169
cork oak 64
corn (maize) *32*, 99, 148, 150,
   182, 184
cornflowers 102–3
corpse flowers 37–8, 79
cortisone 162, *163*
cotton 169, 182, 184
covalent bonds 13–15
   see also conjugated double
      bonds

coordinate covalent
   bonds 34
double and triple bonds 15,
   34, 42, 62–3
fibers to dyes 172
water 16
cow parsnip 87
creosote bush *119*, 120, 133
crocetin *173*, 175
Crocosmia cultivars 55
Crocus sativus 152, 171, 175
croton 129, 131
crystals (raphides) 125
Curcuma longa 153
cuticles 64
cyanobacteria 28–9, 96
cyanogenic glycosides
   51, 133–5
Cynodon dactylon 32
Cyperus papyrus 170
cytosine 71, *72*

Dactylopius coccus 173
daffodils 55
dahlias (*Dahlia* spp.) 42–4
daisies 100
Danaus plexippus 139
dandelions 128–9, 156
Darlingtonia californica 5
Datura spp. *141*, 142
deadly nightshade *3*, 140, *158*
defensive compounds
   see also poisons; toxins
   alkaloids 140–2, 164
   allelopaths 132–3
   cyanogenic and cardiac
      glycosides 133–5, 137–9
   defensive odors 116–19,
      121–3, 153
   defensive tastes 121–5,
      126–8, 153

gums resins and saps 119–21, 128–30
hydrocarbon protectants 43, *44*
irritants 130–2
as pharmaceuticals 157
psychoactive compounds 164
radiation protection 100
saponins 135–7
species susceptibility 115, 122–3
structural defenses 114, *115*, 125, 167
use by insects 139–40
deficiency diseases 149, 150, 152
denaturation 70
desert agave *115*
desert poppies 100
deserts 26, 117, 120, 133
Dianthus caryophyllus 85
Dieffenbachia sp. 125
Digitalis spp. 160
 D. purpurea 137–8
digoxin 138, 160
dimethyl ether *40*
dimethyl sulfide *87*, 88
Dionaea muscipula 5
Dioscorea sp. *161*, 162, *163*
diosgenin 162, *163*
DNA (deoxyribonucleic acid) 71–3, 151, 181–2
dopamine 167
dumb cane 125
durian fruit (*Durio* spp.) *87*, 88
dyestuffs 170–6
 pigments distinguished from 172
 synthetic dyes 175–6
 vat dyes *173*, *174*, 175
dynamic equilibria 20–1

Echinocactus polycephalus 32
Elaeis guineensis 147, *149*
electromagnetic spectrum 34, 35
electron-dot pictures 12–14
electrons 9–12
 photosynthesis 30–1
 radicals 151
 valence electrons 12, 13–14
elements 3–15
 atomic structure and 9–13
 essential elements for humans 9
 essential elements for plants 3–8
 in organic chemistry 39–40
Elodea water plants 20
enantiomers 51–2, 67, *68*, 118, 141
endergonic/endothermic reactions 27
energy
 see also respiration
 chemical reactions 27
 fuels 44
energy levels, electrons 10–12
energy storage
 fats and oils 59
 and human diet 146–7
 starch for 49, 54
English yew 161
enzymes
 acetyl coenzyme A 74–5, *76*
 cofactors 150
 digesting cellulose 56
 and metabolism 28
 as proteins 66
 restriction enzymes 182, *183*
Ephedra spp. 166
ephedrine 164, 166
Epilobium spp. 8

Equisetum spp. 56, 58–9
ergot fungus 167
Ericameria nauseosa 129
Eriodictyon californicum 63, *64*
Erythroxylum coca 126, 165
Eschscholzia californica 97
essential amino acids 67, 148
essential elements 3
essential fatty acids 147
essential oils 80, 88, 117, 177, 179
esters
   pleasant-smelling 85–6, 180
   triglyceride 59–61
ethanol (alcohol) *40*, 41, 164
ethene 42–3
ethyl 2-methylbutyrate 86
ethylene (ethene) 42
eucalyptol 117–18
Eucalyptus spp. 118, 133
   E. globulus 94, *117*
eugenol 84–5, 154–5, 178
Euphorbia spp. 128–31
   E. pulcherrima 129, 130
European yew 115–16
evening primrose 102
exergonic/exothermic reactions 27
expression of oils 177

Fabaceae 134, 140
fall coloration *97*, 98, 99, 102–3, 110–11
fat-solubility
   flavors 155
   odors 179
   pigments 95, 99, 112
   saponins 137
   vitamins 149, 151–2

fats and oils
   cis and *trans* fats 63, 147–8
   lipids 59–65
   palm oil 147, 152
fatty acid biosynthesis 74
fermentation 38–9
fertilizers 8–9, 85
feverfew 156
fiber (dietary) 147
fibers (for rope and textiles) 168–70
fire followers 8
fireweed 8
flavones 100–2
flavonoids 100, *101*, 127
   anthocyanins as 103, 105
   tannins and 109, *110*
flavonols 100
flavors *see* tastes
flax 169
fly agaric *106*, 107
fly and beetle pollinators 86–7, 108
food *see* nutrients
formic acid 131
four o'clock family (Nyctaginaceae) 107
foxglove 137, *138*, 139, 160
fragrances *see* odors; perfumes
frankincense 176–7, *178*
'freebasing' 165
freezing 15, 17, 54, 181
fructose 49–50, *51*, 53
fruits
   acids in unripe 124
   coatings 63
   esters 85
   pits of 133–4
   sugars 49–50
fuchsia 107

functional groups 41, 80–1
  anthocyanins 103
  malodorous compounds 86
  perfumery 180
  terpenes 83, 99
functions of odors 79–80
fungi
  alkaloids in 141, 167
  betalains in 106
  defenses against 114–15, 127

galactose 50, *51*, 53
gallic acid 109, *110*
garlic 122, *123*
gas chromatography (GC) 89, 180
genetic modification 181–4
genetics and nucleic acids 70, 73
Gentiana lutea 126
geraniol 80, 82, *83*, 89, 178
geranium 103
giant corpse flower 37–8
ginger 153, *154*, 155, 156
gingerol *154*, 155
Gleditsia triacanthos 111
glucose
  see also sugars
  asymmetry 52
  as carbohydrate 49–52
  cellulose and 56, *57*
  starch and 54–5
glucosinolates 121–2, *123*
glycine 67, *68*
glycolysis 37, *75*
glycosides 51, 54
  aglycons 128, *136*, 175
  anthocyanins as 103
  cardiac glycosides 51, 116, 130, 136–9, 160

cyanogenic glycosides 51, 133–5
flavonoid 127
products occurring as 85, 132, 154, 160, 174
saponins as 136
soluble pigments as 95, 102–3, *104*
α–glycosidic linkages 54, *55*
β–glycosidic linkages 56, *57*
golden aspen 97
goosefoot family 107
Gossypium spp. 169
grapefruit 127
grapevine *3*, *50*, 103
grasses 32
grayanotoxins *53*, 54
growth inhibition 132–3
guanine 71, *72*
gums and resins 119–21, 130

hallucinogens 163–5, 167–8
'head-space analysis' 88–9
hecogenin 136
Helianthus annuus 60
helium 9, *10*
hemlocks 3
  poison hemlock 140–1
hemoglobin 97
hemp 169
heptane 42–3
Heracleum lanatum 87
herbal medicines 156, 166
herbs 117, 152–5, 156
heroin 165
Heuchera spp. 126–7
Hevea brasiliensis 129
hexane 177
  1-amino- 87
histamine 131

honey locusts *111*
honeybees *see* bees
hop plants 128
horse tails *56*, 58–9
hot climates *see* deserts; water loss
human diet *see* nutrition (human)
humic acids 24
humulone *127*, 128
Humulus lupulus 128
hydrangea 104–5
hydrocarbons 41–4, 63, 86, 99, 179
  see also terpenes
hydrogen bonds
  in nucleic acids 72
  plant fibers 170, 172
  plant pigments 95
  in proteins 69–70, 127
  with sugars 50–1
  in water 16, 17, 19
hydrogen cyanide 133–4
hydrogen ions
  see also acidity
  Calvin Cycle 31
  pH scale 21, 124
hydrogen molecule 14–15
hydrolyzable tannins 109–10

indicaxanthin *106*
indigo *(Indigofera tinctoria)* 171, *173*, 174–5
indole nucleus *166*, 167, *173*, 174
infrared radiation *35*, 96
insecticides 142, 164, 184
insectivorous plants 5, 6
insects
  antinutrients 127
  cardiac glycosides 134, 139–40

cochineal from *171*, 173
genetic modification and 182, 184
herbs as repellents *117*
latex and 128–30
odors as repellents 121
resins and 119
ions
  amino acids as 67
  anions 7, 13
  cations 13, 23–4
  chemical bonding and 13–14
  dissolution 17, *19*
  macronutrients as 6, 7
iron
  as a micronutrient *4*, 7, 24, 26, 125
  as a mordant 172
  in pigments 94, 97, 110
irritants 130–2
Isatis tinctoria 171
isomers *40*, 41–2, 50
  cis and *trans* fats *62*, 63, 147–8
  stereoisomers 52–3, 62–3
isopentenyl pyrophosphate 74, *75*, 84
isoprene rule 84
isoprene units
  in dyestuffs 175
  in odors and perfumes 91, 117, 120, 179
  in taste compounds 128, 154
  terpenes 81–4, 98, 117, 120, 132
  in tetrahydrocannabinol 168
  in vitamins and provitamins 98, 152
isoprenoids *see* rubber; terpenes
isotopes 9–10

Jeffrey pines 42, *44*
jimson weed 142
Juglans nigra 132
juglone 132
jute 169

kapok 169

lactic acid *40*, 41
lactic acid fermentation 39
Lactuca spp. 128
Lapageria rosea 59
Larrea tridentata 119, 120
latex 128–31, *138*, 158–9, 165
Lathyrus odoratus 81, *82*
lavender (*Lavandula angustifolia*) *178*, 179
LD_{50} measure 160, 165
legumes 5, 41, 134, 148
Lemon balm *117*
lemons 81, *82*, 123
  see also *Citrus* spp.
  limonene in 42, *44*, 117
lettuce 128
Lewisia rediviva 126
light reactions 30
lignins 57–9, 170
lima beans 134
limonene 42–3, 81–2, *83*, 117, 178
linalool 179
line structures 40–3
  chlorophyll 33–4
  fatty acids 59, *60*
  isoprene and terpenes 81–3
  oxalin acid *124*
  water *16*
linen 169–70
linoleic acid 59, *61*, 62–3, 147
Linum usitatissimum 169
lipids 59–65

lipines (*Lupinus nanus*) 103–4
Liquidambar styraciflua 111
Lophophora williamsii 167
lutein *98*, 99
luteolin 100, *101*
lycopene *98*, 99
lysergic acid/LSD *166*, 167
lysine *68*, 148

macronutrients 4, 5–6, 8
madder plant 171–2
magnesium
  in chlorophylls 34, 95, 97
  as a macronutrient 6
magnolia 87
maize (corn) *32*, 99, 148, 150, 182
mala mujer 130–1
malic acid 124
Malus spp. 124
malvidin 103, *104*, 105
manganese 7
Manihot esculenta 134
Manilkara spp. 130
mannose 50, *51*, 52, 53
manzanita 53
Marah spp. 137
marijuana 163, 168
mass spectrometry (MS) 89, 180
mauveine 175
medicinal plants 153, 155–63
Mediterranean climates 55–6
Melissa officinalis 117
menthol 117–18, 178
mercury 157
mescaline *166*, 167
metabolism
  enzymes and 28
  primary metabolites 48–9, 76, 146–8
  reaction pathways 74–6

metabolism (*continued*)
  secondary metabolites 81,
    105, 121, 128
  vitamins in 150
metals
  as macronutrients 5
  as micronutrients 7
  as mordants 172
  in pigments 94, 105, 109–10
  tolerance 6–7
methane 14–15, 39–41, 43
methionine *68*, 148
micelles 23–4
micronutrients 4, 7
milkweed 128, 130, 137–9
milkweed beetles 139
mint family 117
molecules
  asymmetric 51–2
  chemical bonding 13–15
  molecular formulae
    34, *40*, 41
  molecular models 42, *44*,
    51–2
  organic structures 39–44
  structural formulae 50–1
molybdenum 7
monarch butterfly 139–40
mono- and disaccharides
  50–1
montbretia *55*
mordants 172–3, 176
morphine 157–9, 164–5
Moschus moschiferus 177
muscone 179
musk deer 177
mustard oils 121
mustard plant 122
Myristica fragrans 152
myrrh 176–7

NAD/NADP1 (nicotinamide
  adenine dinucleotide/
  phosphate) 31, 150
naming of plants 13, 156,
  171, 189
narcissus 100
naringin 127
nasturtiums 97
NDGA (nordihydroguaiaretic
  acid) *119*, *120*, 121, 133
nectars 49–50, 53–4, 102
Nereocystis sp. *29*
Nerium oleander 137, *138*
neurotransmitters 131, 164–6
neutrons 9
Nicotiana spp. 140–1, 165
nicotine 140–1, 164–5
nicotinic acid (niacin) 150
nightshade family 140, 165
Nigritella rubra 85
nitrogen
  see also alkaloids
  as a macronutrient 5, 8
  triple bond in 15
nitrogen-fixing bacteria 5–6
Nopal cactus *64*
norepinephrine
  (noradrenaline) 166, *167*
norethindrone 162, *163*
nucleic acids 70–4, 181
  DNA (deoxyribonucleic
    acid) 71–3, 151, 181–2
  RNA (ribonucleic acid) 71–2
nucleotides 71–2, 74
nutmeg 152, *153*, 154
nutrients (plant)
  elemental 3–4
  excess 8–9
  extraction from soils 7, 13,
    17, 23–7

nutrition (human)
  brassicas 122
  carotenoids in 99–100
  cis and *trans* fats 63, 147–8
  compared with plant 9, 66–7
  essential amino acids 67, 148
  flavorings 152–5
  genetic modification and 181–2
  plant foods in 54–5, 59, 66, 146–9, 152
  vitamins in 148–55
Nyctaginaceae 107
Nymphaea cultivar *20*

oak trees 64, 108
odors
  analysis of 88–90
  aromatic compounds 84–5
  chirality and 83
  defensive odors 116–19
  esters 85–6
  malodorous amines and sulfur compounds 86–8
  perfumes for people 176–81
  terpenes 81–4
Oenothera sp. *102*
oil palm 147, *149*, 152
oils *see* essential oils; fats and oils
Olea europaea 60
oleander 137, *138*, 139
oleic acid 59, *61*, 62–3
olive 59, *60*, 62
onions 7, 122–3
opium/opium poppy 157–9, 165
Opuntia sp. *64*, 106, 173
oral contraceptives 162, *163*
orange milkweed *139*
orbitals, electron 12, 14
orchids *36*, 81, 84

organic chemistry
  aromatic compounds 57
  elemental composition 4–5
  molecular structures 39–44
Oryza sativa 149
osmotic balance 25–6
oxalic acid 124–5
Oxalis spp. 124
  O. pes-caprae 100
oxidation-reduction reactions 28, 30, 150
oxygen
  see also respiration
  double bond 15
  isotopes 10
  role in photosynthesis 5, 28–9
ozone 30

Pacific yew tree 161, *162*
paclitaxel 161, *162*
palm fiber 169
palm oil 147, 152
palmitic acid 59, *60–1*, 62
Papaver somniferum 158, 165
Papaveraceae 165
paper making 170
paprika 153, *154*, 155
papyrus plant 170
Passiflora edulis 134
passion fruit 134
pea family 134, 140
peanuts 59, *60*
pelargon(id)in 103, *104*, 105
pentadecane 86, *87*
peptides 68–70
perfumes, plant *see* odors
perfumes for people 176–81
periodic table 3, 5–6, 9, 11
Perkin, Sir William Henry 159, 175

persimmon tree 126
peyote 163, 167
pH scale 21, 124
   see also acidity;
     hydrogen ions
pharmaceuticals 156–7
Phaseolus lunatus 134
phenolic compounds
   defensive compounds
     118–19, 120–1, 131–3
   flavonoids 101
   flavours 155
   tannins 109, 127
   tetrahydrocannabinol 168
phenyl propanoid units 57
phenylalanine *68*
philodendrons 125
phloem 19
phorbol *131*, 132
phosphoenolpyruvate (PEP) 74–5
phospholipids 65
phosphorus 4–5, 8, 40
phosphorylation 74
photosynthesis 28–36
   see also chlorophyll
   general equation 27
   light reactions and the Calvin Cycle 30
   reaction pathways and 74–6, 150
   respiration as its reverse 37
   role of carbon dioxide 7
   role of oxygen 5, 29, 31
   role of water 22–3, 30–1
phytoene *98*, 99
pickleweed *26*
picrocrocin 154, 175
Pieris spp. 122

pigments
   see also chlorophyll
   accessory pigments 29, 31
   anthocyanins 102–5
   betalains 105–7
   capsanthin *154*, 155
   carotenoids 97–100
   dyestuffs distinguished from 172
   electron energy levels and 11, 34
   fall coloration 110–11
   flavones 100–2
   pollinator specificity 107–8
   tannins 108–10
α-pinene 117–18, 154, 178
pines 108–9, 117–20, 177
   pine resin 42–4, 63, 119–20
pink fireweed *8*
Pinus jeffreyi 42, *44*
Pinus longaeva 109, *119*
Piper nigrum 152
plasmids 182–3
poinsettias *129*, 130
poison hemlock 140–1
poison ivy 131
poison oak 131
poisons 157–8
   see also defensive compounds; toxins
polarity
   gums 119
   water 16, 17
pollen colors 97, *100*, 107
pollinators
   Arum family 38
   color attractants 101–2, 107–8
   nectar attractants 49–50, 53–4
   odor attractants 79, 86
   specific targeting 107–8

polymers 48, 54
  cellulose 56–9, 170
  lignins 57–9, 170
  nucleic acids as 71
  plant fibers 170
  proteins as 65
  rubber 84, 129
  starch 54–5, 147
poppy family 165
Populus tremuloides 97
porphyrins 34, 95, 97
Portulacaceae 107
potassium
  ions 13, 23
  as a macronutrient *4*, 6, 8
potato plant *55*, 140, 182
prickly pear cactus 106, *171*, 173
primary metabolites 48–9, 76, 146–8
progesterone 162, *163*
Protea cordata 53
proteins
  see also enzymes
  primary structure 69, *70*, 71
  secondary, tertiary and quaternary structure 69–70
protons 9–11, 13
Prunella vulgaris 156
Prunus dulcis 134
Prunus lyonii 133, 134
psychoactive plants 163–8
Pultanea pedunculata 5
purple sage 133
putrescin 86, *87*

quercetin 100, *101*
Quercus agrifolia 109
Quercus lobata 109
Quercus suber 64

quinine 126, 128, 157, 159, 175
  structure of *127*, 128, 142, *158*, 159

racemic mixtures 118
radiation protection 100
radicals 151, 185
radioactivity 10, 180–1
radiocarbon dating 10, 121
radish *18*, 102, *103*
Rafflesia arnoldii 79, *80*, 86
rain, acidity 21
rain forests 7–8, *25*, 163
ramie 169
Raphanus sativus 103
reaction pathways 74–6, 81
reactive oxygen species (ROS) 151–2, 185
recombinant DNA 182–3
red cabbage 102–3, 105
resins 42, 63, 119–21, 177
respiration 36–9
  root hairs 24
  water plants 19–20
retinol 151
Rheum rhabarbarum 23
Rhododendron spp. 53
Rhoeo sp. *18*
rhubarb *23*, 124
rice 38
ricin 66
Ricinus communis 66
ripening hormones 42
RNA (ribonucleic acid) 71–2
root hairs 17, *18*, 24
root nodules 5–6
Rosa californica 94
Rosa rugosa 3
Rosaceae 134
rose fragrance 79–80, 89, *90*

rosemary 117–18, 153–4
rosin 120
Rosmarinus officinalis 118, 153
rubber 84, 129
rubber rabbitbrush 129
Rubia tinctorum 171–2
Rubus parviflorus 29
ruminants 56
Ruzicka, L 84
rye 167

Saccharum spp. *50*
saffron 152–5, 171, 175
sage 156
saliclin *160*
Salicornia sp. *26*
salicylic acid 124
    acetyl- 160
Salix spp. 124, 160
salts 13, 25–7
Salvia leucophylla 133
Salvia officinalis 156
sapodilla tree 130
Saponaria spp. 135
saponins 135–7, 142, 162, *163*
saps *see* latex
Sarcodes sanguinea 36
saturated fats 61–2
scale insects *171*, 173
scents *see* odors; perfumes
Scirpus acutus 20
scopolamine *141*, 142, 166
scouring rush *56*, 58–9
Secale cereale 167
secondary metabolites 81, 105, 121, 128, 141
seeds
    dyes and perfumes from 175, 177
    in human nutrition 59, 66, 148–50, 152
    spread by animals 85, 88, 116
selenium 6, 9
self-heal 156
Sequoia sempervirens 56, *126*, 127
serotonin 131, *167*
serpentine 6–7
showy milkweed *138*
sinapyl alcohol 57, *58*
sinigrin 121, *123*
sisal 169
small alpine orchid *85*
smells *see* odors
snow plant *36*
soap plant, soap tree and soapwort 135
soils
    acidity 24–5, *26*, 109
    nutrient extraction from 7, 13, 17, 23–7
    water retention and 23–4
Solanaceae 140, 165
solanine 140, *141*, 142
Solanum lycopersicum 97
Solanum tuberosum 55, 140
solubility *see* fat-solubility; water solubility
sorghum (*Sorghum* spp.) 134
sorrel 124
sourgrass *100*, 124
soybeans 182, 184
space-filling models 16, 33–4, 59, 61
spices 84, 152–5, 175
spinach 124
spurges 129
starch 54–5, 147

Subject Index

steam-distillation 88, 120, 177
stereoisomers 52–3, 62–3
　see also chirality
　terpenes 83, 99, 118, 129
steroids
　alkaloids or 142
　cardenolides as 136,
　　138, 160
　oral contraceptive 162, *163*
　plant-derived 162, *163*
　saponins as terpenes or 136,
　　142, 162
stinging nettle 130–1
stinking milk vetch 6, 7
stomata 17, *18*, 31–3
stromatolites 96
structural formulae 50–1, 124
structural polymers *see*
　cellulose; lignins
suberin 64
sugar beet 182
sugar cane *50*
sugars
　see also carbohydrates
　glycolysis 37
　isomers 50–1
　from photosynthesis 31
　solution 19, 49–50
sulfurous compounds 86, 121–3
sunflowers *56, 60*
surface-active compounds
　135, 137
surface tension 15
sweet pea 81, *82*
sweetgum 111
sweetness 49, 53
Syzygium aromaticum 84, 152

Tanacetum parthenium 156
tannins 108–10, 111, 126–7

Taraxacum officinale 156
Taraxum spp. 128, *129*
Taro plant 125–6
tastes
　defensive compounds
　　121–3
　herbs and spices 152–5
taxol 161, *162*
Taxus baccata 115–16, 161
Taxus brevifolia 161, *162*
terpenes 81–4, 142
　carotenoids as 84, 98, 152
　crocetin and other dyes
　　173, 175
　defensive 117–18, 133
　diterpenes 120, *131*, 132
　monoterpenes and
　　flavor 154–5
　monoterpenes and odor 83,
　　117–18, 120, 175, 178
　saponins as 136
　sesquiterpenes 179
　steroids derived from 136
　THC 168
　triterpenes 136
tetrahydrocannabinol
　(THC) 165, *168*
Tetraopes tetrophthalmus 139
thiamine 149–50
thimbleberry 29
thiopropanal-S-oxide *123*
thorn-apple 142
thymine 71, *72*
tobacco plant 140, 165
α-tocopherol *149*, 151–2
tomatoes 97
total synthesis 159
Toxicodendron
　diversilobum 131
Toxicodendron radicans 131

toxins
  see also poisons
  cyanogenic glycosides 133–5
  grayanotoxins 53, 54
  LD_{50} measure 160, 165
  species susceptibility 115,
    137, 141–2
  toxic nectars 53–4
  toxic proteins 66
trans fats 63, 147–8
Trichocereus grandiflorus 94
Trifolium repens 133, 134
triglycerides 59–62
1,2,4-trithiolane,
  3,5-dimethyl- 87
tropane ring 166
tubers 54–5
tule reeds 20
tulips 53, 55, 104
turmeric 153
turpentine 81, 120
Tyrian purple 173–4

ultraviolet light 35
  absorption 98, 99–102, 107
  insect pollinators 107
  potential damage 30, 151
unsaturated fats 61, 62, 147
Urtica dioica 130–1
urushiols 131
uses of plants
  dyestuffs 170–6
  fibers 168–70
  genetic modification 181–4
  human nutrition 146–55
  medicinal uses 153, 155–63
  oxygen replenishment 145
  perfumery 176–81
  psychoactivity 163–8
  species names and 156, 171

vacuoles 95, 102–3, 105, 107
valence electrons 12, 13–14
Vanilla planifolia 84, 85
vanillin 81, 84–5, 154–5, 178
Venus Flytrap 5
viola spp. 11
vitamin A 151
  β-carotene as precursor
    99, 151
vitamin C 149, 150–1
vitamin E 149, 151–2
vitamins 148–55
  deficiency diseases 149,
    150, 152
  excess 150
  fat-soluble 151–2
  water-soluble 149–50
vitamins B_1, B_2, B_3 149–50
Vitis spp. 3, 50, 103
Viverra civetta 177
volatile mixtures
  analysis 89–90, 180
  Arum family 37–8
  defensive systems 118,
    120–3, 133
  flavor components 155
  odors and perfumes 86–7,
    89, 91, 124, 178–80

Wallach, O 83
water
  contribution to elemental
    composition 4
  density as liquid 15, 17
  dissociation 21
  retention and soil type
    23–4
  role in photosynthesis 22–3,
    30–1
  unusual properties 15–23

water loss
   adapted photosynthesis 32
   protective barriers 42, 63,
      114, 118, 120
water plants 19–20
water repellents 63–4
water solubility 17–19
   plant pigments 95, 102–3,
      104
   sugars 49–51
wavelengths 34, *35*, 94
   anthocyanins 103, 105
   betalains 105–6
   chlorophylls and 34, *35*, 96
   flavonoids and
      carotenoids 99, 102, 105–6
waxes 63
wedge-and-dash pictures 12–14,
   51–2, 118, 162
white clover *133*, 134
wild cucumber 137
wild onion 7

wild rose *3*
wild yam *161*, 162, *163*
wildfires 8, 118
willow 124, 160
woad 171
wolf's milk 128–9
Woodward, Robert 159
Wyethia mollis 56

xanthophylls *98*, 99–100
xylem 17, *18*

yellow gentian 126
Yerba Santa 63, *64*
yuccas 135, 169–70
   Y. elata 135

Zantedeschia aethiopica 125
Zea mays *32*, 99
zeaxanthin 99
zinc 4, 7
Zingiber officinale 153, 156